Biodiversity and Health

Series Editor
Françoise Gaill

Biodiversity and Health

Linking Life, Ecosystems and Societies

Serge Morand
Claire Lajaunie

ISTE PRESS

ELSEVIER

First published 2018 in Great Britain and the United States by ISTE Press Ltd and Elsevier Ltd

ISTE Press Ltd
27-37 St George's Road
London SW19 4EU
UK

www.iste.co.uk

Elsevier Ltd
The Boulevard, Langford Lane
Kidlington, Oxford, OX5 1GB
UK

www.elsevier.com

British Library Cataloguing-in-Publication Data
A CIP record for this book is available from the British Library
Library of Congress Cataloging in Publication Data
A catalog record for this book is available from the Library of Congress
ISBN 978-1-78548-115-4

Contents

Foreword

Global population growth, industrialization and geopolitical problems accelerate global changes and are in turn leading to a significant erosion of biodiversity, significant degradation of ecosystems and large migratory movements, both by human and animal populations.

Due to an increased rate of globalization and various global changes, we are now facing an increase in the spread of infectious agents, rising risks of pandemics and increased incidence of many chronic non-infectious diseases. These changes highlight the globalization of health risks, as well as the importance of the human-animal-ecosystem interface in the evolution and emergence of pathogens as well as in the determinism of chronic diseases. These health crises outline the interdependence between human, animal and ecosystem health.

The "One World–One Health" concept emerged in 2004 during a conference of the Wildlife Conservation Society. The concept aims to combine biodiversity conservation with public health objectives by making the connection between the emergence of infectious diseases stemming from wildlife and damage to biodiversity. This new concept allows for an interdisciplinary approach towards health problems by bringing together ecologists, physicians, anthropologists, biologists, demographers, among others, in order to address the etiology and prevention of chronic non-communicable diseases and infectious diseases in an essential and innovative way.

This book demonstrates the diversity of these scientific approaches at the heart of the relationship between biodiversity and health, by first exploring the major trends in the spread of infectious diseases and its links with animal biological diversity and human cultural diversity. In this way, it is essential to have knowledge of the ecosystem dynamics which drive the processes, leading to the appearance or resurgence of infectious agents, their transmission, and their extinction in the natural world, as this knowledge in turn leads to an understanding of infectious risks. In the second stage of the approach, the causes and mechanisms, which explain this animal biological diversity, as well as human biological and cultural diversity, are to be found in human evolution and history, from the first case of human migration out of Africa, to the emergence of agrarian civilizations and the first cases of globalization.

It is now known that in most cases, the total eradication of infectious diseases is impossible since there is a permanent coevolution between the resistance of disease vectors and the fight against them. From this particular observation, evolutionary medicine, otherwise known as Darwinian medicine, came to fruition. This approach aims to understand the ecological and evolutionary principles at the origin of many diseases in order to improve treatment and prevention. The hypotheses derived from evolutionary theory demonstrate, for example, the importance of preventing or slowing down the evolution of resistance or the appearance of unintended effects. Theory, modeling and empirical experiments are therefore crucial for classical management methods and new fine manipulation methods in the fight against diseases.

These transdisciplinary approaches to biodiversity, well-being and health also highlight other important aspects: the importance of scientific expertise, but also the importance of ethical and legal, governance and economic issues.

Beyond this fundamental knowledge, the biodiversity/health approach provides scientists with an extremely valuable source of inspiration. We can take the example of the re-appropriation of biocultural knowledge embodied in traditional pharmacopoeia, knowing that the loss of biodiversity results in the loss of chemodiversity in secondary metabolites essential for interactions among plants, animals and humans, such as metabolites with anti-parasitic and anti-microbial properties. On the subject of ecosystem services rendered by nature, this knowledge allows us to create a link between biodiversity and human well-being. In a more innovative way, new technologies such as CRISPR/Cas system gene editing and manipulation should be able to help us in the fight against certain vectors, including malaria for example, but it will be necessary to have a better understanding of the mechanisms and target specificity, their consequences at the population level and the adaptive mechanisms that could lead to unintended consequences.

This book provides an in-depth and fascinating discussion of the many aspects of the close relationship between biodiversity, health and human well-being.

Martine HOSSAERT
Institute of Ecology and Environment
CNRS

Introduction

In a 2017 report on human rights obligations related to the preservation and sustainable use of biological diversity, United Nations special rapporteur J.H. Knox[1] stressed that the relationship between human rights and biodiversity is not well understood. In principal:

> "The full enjoyment of human rights, including the rights to life, health, food and water, depends on the services provided by ecosystems. The provision of ecosystem services depends on the health and sustainability of ecosystems, which in turn depend on biodiversity".

It is therefore necessary to learn more about the complex relationship between living organisms, ecosystems and societies, subjects that this book proposes studying, including the numerous interactions between biodiversity and health. The book offers a broad and integrated view of the scientific disciplines which contribute to health studies and it illustrates how ecological, environmental, medical or even how social sciences can contribute to improving human health through the conservation of biodiversity and the services that biodiversity provides to animal and human societies.

By presenting a brief history of the relationship between health and biodiversity, Chapter 1 demonstrates how international organizations, large

---

1 Knox is a special rapporteur on the question of human rights obligations pertaining to the provision of a clean, healthy and sustainable environment. The extract used is from: http://ap.ohchr.org/documents/dpage_e.aspx?si=A/HRC/34/49 (Document E).

conventions and other international forums dealing with the environment, biodiversity, agriculture, development, animal health and human health appropriated the concepts developed by Millennium Ecosystem Assessment [MA 05] and also proposed causal links between biodiversity conservation, the functioning of ecosystems and the quality of ecosystem services for health and human well-being. However, this book outlines that these relationships have been previously identified, and identified in a way that is still relevant and consistent with the One Health approach.

Chapter 2 focuses on the relationships between biological diversity, cultural diversity and infectious diseases, outlining the different correlations that exist between biodiversity and pathogens and between pathogen diversity and cultural diversity. Animal domestication has been accompanied by an increase and sharing of infectious diseases between pets and humans. Thus, infectious diseases play an important role in the understanding of the coevolution between nature and culture.

Chapter 3 focuses on the relationship between the loss of biodiversity and the emergence of infectious diseases; this is notably highlighted through epidemiology, the study of zoonoses, mechanisms of emergence and genetic diversity. Cited scientific works demonstrate that the loss of biodiversity is accompanied by an increased risk in the emergence of new infectious diseases. In addition, Chapter 4 focuses on the loss of biodiversity and the emergence of non-infectious diseases through the study of host-parasite coevolution, the evolution of the immune system under pressure from parasite diversity, or even from hygiene hypothesis or that of the emergence of atopic sensitization.

The growth of the human population is accompanied by an increase in overall energy consumption, urbanization, development of road infrastructure and even agronomic intensification. Chapter 5 tackles the topic of anthropogenic stressors in relation to these transformations, in particular the effects of phytosanitary and biocidal products and the psychological impacts and responses to environmental stressors (eco-physiology stress) when endocrine disrupters are involved.

Living beings adapt to the planet's new environmental conditions and Chapter 6 considers the response of biodiversity to anthropization. This chapter presents information on the different forms of adaptation: behavioral adaptation such as synanthropy, or genetic adaptation such as the resistance of organisms targeted by insecticides, anti-parasitic drugs, and antibiotics as well as resistance to genetically modified plants. Chapter 6 presents the mechanisms and consequences of the evolution of virulence and resistance to new biotechnologies as well as the role of ecological and evolutionary engineering.

Biodiversity is often presented as a principal source of pharmaceutical medicine and molecules for the agro-chemical industry, and Chapter 7 focuses on animal and human pharmacopoeias, notably by examining the evolutionary origin of self-medication in animals and hominids, the diversity of secondary plant metabolites and ethnobotany and traditional medicine. Chapter 7 also details bioprospecting and the necessity of access and benefit sharing, non-compliance and biopiracy, placing importance on traditional pharmacopoeia and highlighting the relationship between loss of biodiversity and loss of traditional knowledge.

Well-being in terms of its objective and subjective elements are detailed in Chapter 8. The psychological impacts of the natural environment, evolutionary psychology and its relationship with well-being (the habitat and prospect refuge theory, topophilia and biophilia) or the importance of traditional knowledge enables us to demonstrate the profound and emotional ties that bind us to the living.

Ecosystem services (Chapter 9), through the concept of ecosystem health, are presented as a way to take into account the relationship between health and biodiversity. This also involves developing the notion of ecosystem services and the need to establish compromise between services, economic development and health.

Prospects and global scenarios for health and biodiversity (Chapter 10) are generally drawn up in broad sectoral areas: climate disturbance, demography, urbanization, globalization, land use, agriculture and livestock and living resources. These domains analyzed separately and sometimes in interaction have independent effects on health and biodiversity and on the nature of biodiversity-health relationships. Worst case scenarios attempt to determine planetary limits in order to maintain a control variable at a "safe" distance from the hazardous threshold and avoid collapse and a swing towards unwanted conditions. This type of scenario has direct implications on state governance, particularly on national security. It is therefore necessary to construct integrated models of society, biodiversity and ecosystem services through observatories of participatory methods (associating users, managers, politicians) but also to introduce experts and appropriate knowledge representation tools.

This raises the question of the governance of biodiversity and health at different levels of decision making (Chapter 11). There is no international body explicitly including both biodiversity and health in its mandate, but issues related to biodiversity and health have themselves been indirectly taken into account by a large number of international organizations. Determining and taking regional issues into account is fundamental to the implementation of internationally defined policies and strategies. International regulation must be effectively implemented at local level, though it remains to be seen how this implementation can be carried out by respecting the underlying defined regulation and in keeping with political choices.

Chapter 12, entitled "Ethics, values and responsibilities", enables us to reflect on the ethical implications of research in areas of biodiversity and health, also notably on scientific pluralism and the type(s) of ethics involved (human, animal, environmental). This chapter also raises the question of ethical dilemmas and opts for a necessary pluralism of ethics.

The issues of law, justice and scientific knowledge in the consideration of health and biodiversity are addressed in Chapter 13: they are addressed by considering the complexity of scientific knowledge and by considering information in political decisions. It is a question of tackling the legal stakes in order to achieve progress in the development of a law in touch with concrete issues, the use of scientific knowledge by citizens for the purpose of collective knowledge, and finally how scientific knowledge can be considered from a human rights angle.

Finally, the conclusion to this book argues for both a socio-ecology of health, for a scientific pluralism that avoids the trap of reductionism and purely techno-scientific approaches. It also argues for an adaptive environmental law.

1

A Brief History on the Links between Health and Biodiversity

1.1. Introduction

The Rockefeller Foundation/Lancet Commission report on planetary health (2015) shows how improving human health has been primarily done to the detriment of the environment. The report highlights that global environmental change (including climate change, deforestation, land use change and loss of biodiversity) is a serious threat to human health. It proposes adopting the concept of planetary health, which is based on recognizing that human health depends on natural systems functioning properly. However, upon first inspection, the definitions of health and biodiversity do not show clear links between the two.

In the Preamble to its 1946 Constitution, the World Health Organization (WHO) defines health as follows:

> "Health is a state of complete physical, mental and social well-being and not merely the absence of disease or infirmity".

For the WHO, therefore, health is not only about disease, it is also and above all a state of well-being, both at an individual and collective level. Earlier in its history, the WHO embraced the importance of the environment and its quality as a contributor to human health and well-being.

The Convention on Biological Diversity (CBD) in 1992 defines biodiversity as:

> "Biological diversity means the variability among living organisms from all sources including, inter alia, terrestrial, marine and other aquatic ecosystems and the ecological complexes of which they are part: this includes diversity within species, between species and of ecosystems".

The links between environmental diversity and health were approached from different angles at the Unesco Biosphere Conference in 1968 which studied the scientific basis for rational use and conservation of biosphere resources. The loss of diversity had already been associated with the deterioration of physical and mental health and the proceedings of the conference stated:

> "Whether the challenges come from physical or social forces, the diversity of environments is of crucial importance for the evolution of man and his societies because the ultimate results of a stereotyped and equalized environment can be and often is an impoverishment of life, a progressive loss of the qualities that we identify with humanness and a weakening of physical and mental health. Our policy should be to preserve or to create as many diversified environments as possible". [UNE 70]

1.2. Millennium Development Goals for Ecosystem Services

Nevertheless, it was not until the Millennium Development Goals and especially the Millennium Ecosystem Assessment [MA 05] that a report on the state of the world's ecosystems finally established a functional link between biodiversity, health and human well-being. This link is expressed through the notion of "ecosystem services". The degradation of ecosystem functioning and associated losses of biodiversity have negative impacts on the quality of ecosystem services, thus affecting the safety, health and welfare of populations. The notion of ecosystem services emerged from the Ecosystem Approach, which is also defined by the CBD as:

> "The ecosystem approach is a strategy for the integrated management of land, water and living resources that promotes conservation and sustainable use in an equitable way. Thus, the application of the ecosystem approach will help to reach a balance of the three objectives of the Convention: conservation; sustainable use; and the fair and equitable sharing of the benefits arising out of the utilization of genetic resources.
>
> An ecosystem approach is based on the application of appropriate scientific methodologies focused on levels of biological organization, which encompass the essential structure, processes, functions and interactions among organisms and their environment. It recognizes that humans with their cultural diversity are an integral component of many ecosystems"[1].

The MA (Millennium Ecosystem Assessment) was implemented in 2001 to address the consequences of global changes of human well-being in ecosystems. This was done to provide the scientific basis for improving conservation and sustainable use of living resources. Over 1,360 global experts worked on the MA (Figure 1.1). A more precise definition of ecosystem services is given by the MA as:

> "Ecosystem services are the benefits people obtain from ecosystems. These include provisioning services such as food and water; regulating services such as flood and disease control; cultural services such as spiritual, recreational, and cultural benefits; and supporting services, such as nutrient cycling, that maintain the conditions for life on Earth".

While any change in biodiversity can influence the contribution of ecosystem services to human well-being, biodiversity conservation is crucial for human development and reduction of poverty. This was therefore the first time that a link was established between the degradation of ecosystems and fight against poverty and its consequences, such as the state of health and well-being of populations.

1 CBD, COP 5, Decision V/2, Ecosystem approach, 15–26 May 2000, Nairobi, Kenya.

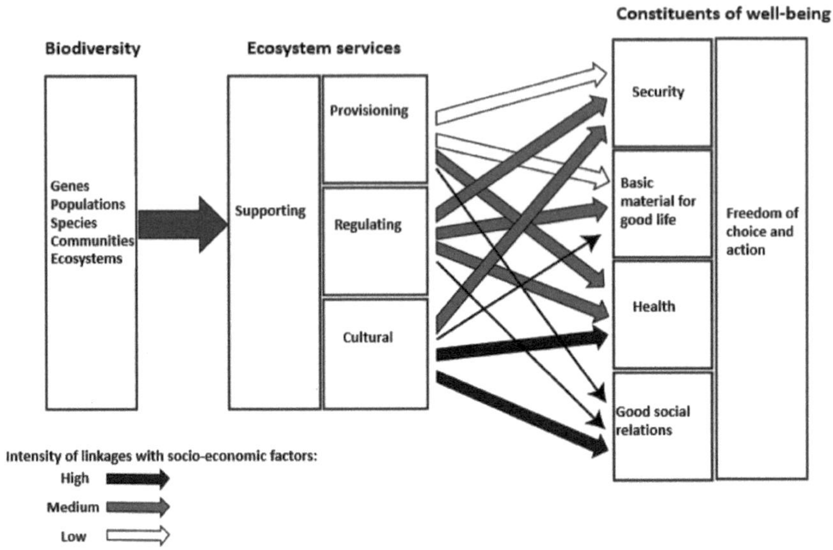

Figure 1.1. *Ecosystem services according to the Millennium Assessment [MA 05]*

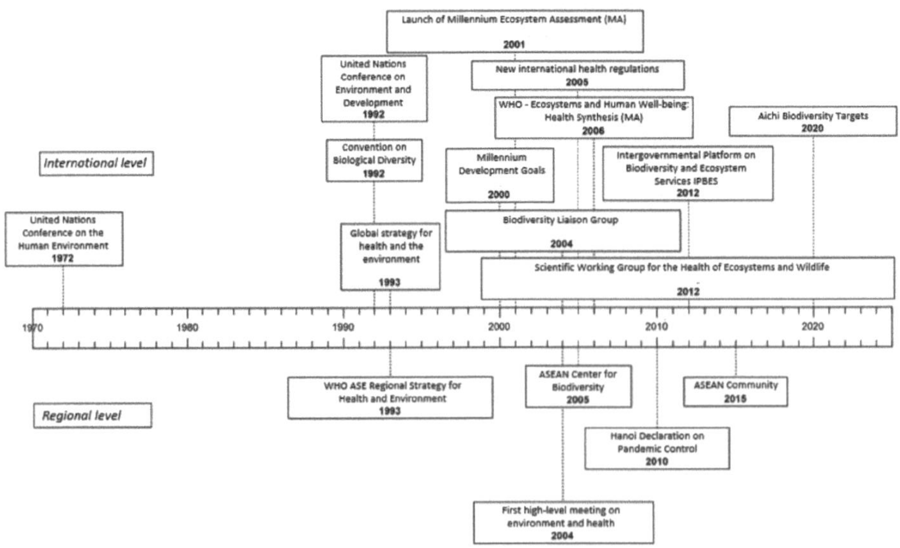

Figure 1.2. *Timeline of the main international conventions and texts relating to the environment, biodiversity and health (from [LAJ 16])*

These international approaches to biodiversity, well-being and health emphasize two aspects: the importance of scientific expertise and the importance of legal, governance and economic issues.

Moving from one conference to another (Figure 1.2, Box 1.1), the need for scientific expertise in biodiversity was increasingly highlighted, particularly at the Paris Conference (2005), which called for international expertise on biodiversity. This is what led to the Intergovernmental Panel on Climate Change (IPCC) (Box 1.1).

International Institutions:

WB, World Bank: http://www.worldbank.org/

Consultative Group on International Agricultural Research (CGIAR): http://www.cgiar.org/

United Nations Foundation: http://www.unfoundation.org/

Global Environment Facility (GEF): www.thegef.org

Intergovernmental Panel on Climate Change (IPCC): http://www.ipcc.ch/

Intergovernmental Science-Policy Platform on Biodiversity and Ecosystem Services (IPBES): http://www.ipbes.net/

International Council for Science (ICSU): http://www.icsu.org/

World Health Organization (WHO): http://www.who.org/

World Organization for Animal Health (OIE): http://www.oie.int

Food and Agriculture Organization of the United Nations (FAO): http://www.fao.org/

United Nations Educational, Scientific and Cultural Organization (UNESCO): http://www.unesco.org/

United Nations Environment Programme (UNEP): http://www.unep.org/

United Nations Development (UNDP): http://www.undp.org/

World Conservation Union (IUCN): http://www.iucn.org/

Conventions and Agreements:

Convention on Biological Diversity: www.cbd.int

Climate Change Adaptation Database from the Convention on Biological Diversity: http://adaptation.cbd.int/

United Nations Framework Convention on Climate Change: http://unfccc.int/2860.php

Ramsar Convention on Wetlands: http://www.ramsar.org

United Nations Convention to Combat Desertification: http://www.unccd.int/

World Heritage Convention: http://whc.unesco.org/en/convention/

Convention on the Conservation of European Wildlife and Natural Habitats: http://www.coe.int/en/web/bern-convention

Convention on the Conservation of Migratory Species of Wild Animals: http://www.cms.int/en/

Convention on International Trade in Endangered Species of Wild Fauna and Flora: https://www.cites.org

Convention on the Protection and Promotion of the Diversity of Cultural Expressions: http://en.unesco.org/creativity/convention

FAO - Commission on Genetic Resources for Food and Agriculture: http://www.fao.org/nr/cgrfa/cgrfa-home/en/

World Trade Organization - Trade-Related Aspects of Intellectual Property Rights (TRIPS): https://www.wto.org/english/tratop_e/trips_e/trips_e.htm

Nagoya Protocol on Access and Benefit-Sharing: https://www.cbd.int/abs/

International Programs, Non-Governmental Organizations:

BirdLife International (BLI): http://www.birdlife.net/

Center for International Forest Research (CIFOR): http://www.cifor.cgiar.org

Conservation International (CI): http://www.conservation.org/

EcoHealth (EH): https://ecohealth.net/en/

The Economics of Ecosystems and Biodiversity (TBEE): http://www.teebweb.org/

Environmental Research Web: http://environmentalresearchweb.org/cws/home

Global Invasive Species Programme (GISP): http://www.issg.org/

Global Ballast Water Management Programme (GWMP): http://archive.iwlearn.net/globallast.imo.org/globallast.imo.org/index.html

Group on Earth Observations Biodiversity Observation Network (GEO BON): http://geobon.org/

Fauna & Flora International (FFI): http://www.fauna-flora.org/

Future Earth: http://www.futureearth.org/

International Council for Game and Wildlife Conservation (CIC): http://www.cic-wildlife.org/

One World, One Health (OWOH): http://www.oneworldonehealth.org/

Wetlands International (WI): http://www.wetlands.org/

The Wildlife Trusts: http://www.wildlifetrust.org/ http://www.wildlifetrusts.org/

World Conservation Society (WCS): http://www.wcs.org/

World Wide Fund for Nature (WWF): http://www.panda.org/

Association for Tropical Biology and Conservation (ATBC): http://www.tropicalbio.org/

International Geosphere-Biosphere Programme (IGBP): http://www.igbp.net/

Box 1.1. *Institutions, organizations, conventions and programs referring to biodiversity, environment, health and ecosystem services*

This request was endorsed by the United Nations Assembly in December 2010 and the Intergovernmental Science-Policy Platform on Biodiversity and Ecosystem Services (IPBES) emerged in 2012 (Figure 1.3).

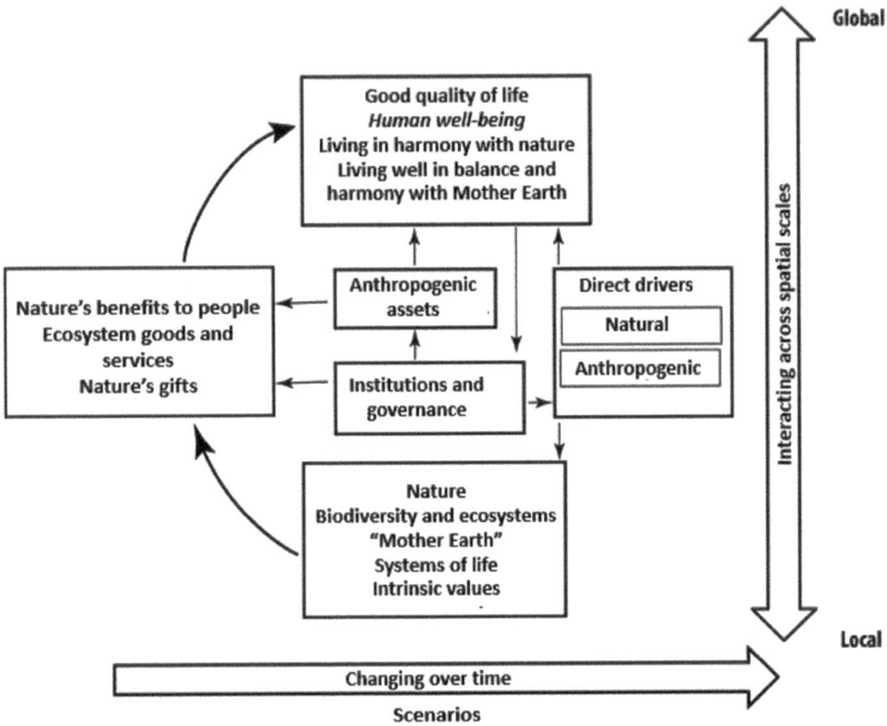

Figure 1.3. *The IPBES conceptual framework (Intergovernmental Science-Policy Platform on Biodiversity and Ecosystem Services). The boxes and arrows denote the elements of nature and society that are at the heart of IPBES. Box titles denote intelligible inclusive categories and are relevant to all stakeholders involved in IPBES, meaning those that encompass the categories of Western science and equivalent categories from other knowledge systems (taken from [DIA 15])*

1.3. From environmental health to "one health"

Although the World Health Organization (WHO) did not always show great interest in biodiversity, this was not the case for the World Organisation for Animal Health (OIE) and the Food and Agriculture Organization (FAO). A reason for this may be the consideration of different themes highlighted by terms that include the word health alongside the words environment, ecology or ecosystem, but for very different domains and varied concepts (Box 1.2).

Environmental health

Environmental health is a branch of public health that addresses all aspects of the broader environment that can affect human health. Environmental health has been defined by the World Health Organization as aspects of human health and diseases that are determined by environmental factors. Environmental health also refers to the assessment and control of environmental factors that can potentially affect health. Environmental health focuses on the direct pathological effects of chemicals, radiation and certain biological agents in dwellings, in urban, agricultural or natural environments, as well as their indirect effects on well-being. It connects the social and cultural environments, as well as genetic components.

Ecological health

Ecological health is used both for human health and to describe the state of the environment. For human health, it is closely related to environmental health as it focuses on exposures of synthetic chemicals, such as pesticides or food packaging residues that are released into the environment. But rather than focusing on the health impacts of environmental factors, it seeks to address the causes. Urbanism has appropriated ecological health in the context of green cities and agriculture in the context of organic plant or animal production (without the use of inputs or synthetic pesticides). Ecological health is different from ecosystem health.

Ecosystem health

Ecosystem health is a metaphor used to describe the state of an ecosystem and originates from the work of Aldo Leopold who spoke of the health of the earth as he was sick of the violent practices and thus presented preservation as the medicine for the earth. This medical metaphor is used as a communication tool and is accompanied by "health" indicators to evaluate desirable environmental objectives [COS 92]. They may involve conservation of species, maintenance of evolutionary and ecological processes or the functioning of ecosystems.

Box 1.2. *Environmental, ecological and ecosystem health*

Karr [KAR 96] distinguished between ecosystem health and ecological health by showing the interactions between ecosystem integrity and health. He stated that biological systems are complex entities with multiple components that require protection, which is not possible according to him. He therefore proposed to define "biological attributes that can be measured with precision to provide reliable information about a biological condition". According to Karr [KAR 96], four characteristics are required: species richness, species composition, individual health and trophic structure (food web). These allow species changes to be detected as well as ecological processes of energy and material flows throughout food webs, and the health of individuals likely to influence demographic processes. A link is therefore made between health (of animals and plants) and ecological function.

The MA with its emphasis on ecosystem services for health regulation has also been actively involved with major non-governmental conservation organizations, particularly the World Conservation Society (WCS).

In 2004, the WCS, which is one of the leading international non-governmental biodiversity organizations (Box 1.1), promoted the "One World One Health" initiative. For the WCS, the aim was to link biodiversity conservation with public health targets by linking the emergence of infectious diseases from wildlife to biodiversity. This initiative gave rise to the "Twelve Manhattan Principles" (Box 1.3), which are the basis of what has become "One Health", an approach supported by major UN organizations like the FAO, the WHO and the OIE.

1) Recognizing the link between human, domestic animal and wildlife health, and the threat disease poses to people, their food supplies and economies, and the biodiversity essential to maintaining a healthy environment and functioning ecosystems we all require.

2) Recognizing that decisions regarding land and water use have real implications for health. Alterations in the resilience of ecosystems and shifts in patterns of disease emergence and spread manifest themselves when we fail to recognize this relationship.

3) Including wildlife health science as an essential component of global disease prevention, surveillance, monitoring, control, and mitigation.

4) Recognizing that human health programs can greatly contribute to conservation efforts.

5) Devising adaptive, holistic and forward-looking approaches to the prevention, surveillance, monitoring, control and mitigation of emerging and resurging diseases that fully account for the complex interconnections among species.

6) Seeking opportunities to fully integrate biodiversity conservation perspectives and human needs (including those related to domestic animal health) when developing solutions to infectious disease threats.

7) Reducing demand for and better regulating international live wildlife and bushmeat trade, not only to protect wildlife populations but to lessen the risks of disease movement, cross species transmission and the development of novel pathogen-host relationships. The costs of this worldwide trade in terms of impacts on public health, agriculture and conservation are enormous, and the global community must address this trade as the real threat to global socioeconomic security.

8) Restricting the mass culling of free-ranging wildlife species for disease control to situations where there is a multidisciplinary international scientific consensus that a wildlife population poses an urgent, significant threat to human health, food security or wildlife health more broadly.

9) Increasing investment in the global human and animal health infrastructure commensurate with the serious nature of emerging and resurging disease threats to people, domestic animals and wildlife. Enhanced capacity for global human and animal health surveillance and for clear, timely information-sharing (that takes language barriers into account) can only help improve coordination of responses among governmental and non-governmental agencies, public and animal health institutions, vaccine/pharmaceutical manufacturers and other stakeholders.

10) Forming collaborative relationships among governments, local people and the private and public (i.e. non-profit) sectors to meet the challenges of global health and biodiversity conservation.

11) Providing adequate resources and support for global wildlife health surveillance networks that exchange disease information with the public health and agricultural animal health communities as part of early warning systems for the emergence and resurgence of disease threats.

12) Investing in educating and raising awareness among people of the world and influencing the policy process to increase recognition that we must better understand the relationships between health and ecosystem integrity to succeed in improving prospects for a healthier planet.

Box 1.3. *The Manhattan Principles "One World, One Health" (http://www.oneworldonehealth.org/)*

The major conventions and other international forums that deal with the environment, biodiversity, agriculture, development, animal health and human health therefore appropriated the concepts developed by the MA [MA 05]. These concepts proposed the existence of causal links between biodiversity conservation, the functioning of ecosystems and the quality of ecosystem services for human health and well-being. But is there any scientific evidence that human health depends on biodiversity [CHI 08]?

1.4. Formerly recognized links

Complex links between the environment, climate, biodiversity and health are recognized and asserted through public health. Thus, Thomas Logan, who was a doctor of the Californian "New Frontier", wrote in 1859:

> "A knowledge of the etiology of diseases can best be attained by studying the affections of different localities in connection with every condition and circumstance calculated to operate prejudicially or otherwise upon the health of the inhabitants. Such philosophical investigation is particularly useful in tracing the modifications diseases may undergo from the agency of causes of a local or special character; and being also calculated to elucidate the relationship of diseases to climate, to the prevailing geological formations—the fauna, the vegetables, the minerals, the waters, which vary with the earth's crust, …". (Thomas Logan, Transactions of the American Medical Association, 1859, quoted in Nash [NAS 06])

Thomas Logan was confronted with the diseases of his fellow citizens of European origins, who were colonizing the "new frontier" habitats (but not including native Amerindian populations), and proposed an environmental geographical approach to human health. Interestingly, Logan wrote his essay at a time when Darwin's theory of evolution (co-presented with Wallace in July 1858) was not yet popularized in North America. On the other hand, Logan was certainly aware of the writings of Alexander Von Humboldt, the founder of modern biogeography (with his "*Essay on the Geography of Plants*", which was published in 1807) who had conducted several important geographical and natural history expeditions in South, Central and North America.

The diagram proposed by Worthington in 1938 (cited in [RAV 06]) is even more fascinating (Figure 1.4). This diagram is one of many scientific works from colonial administrations which were coordinated by scientific networks on the development of European colonies (agriculture, forestry, fisheries and health). It is also in the European colonies that the origin of geography of health must be sought (Box 1.4).

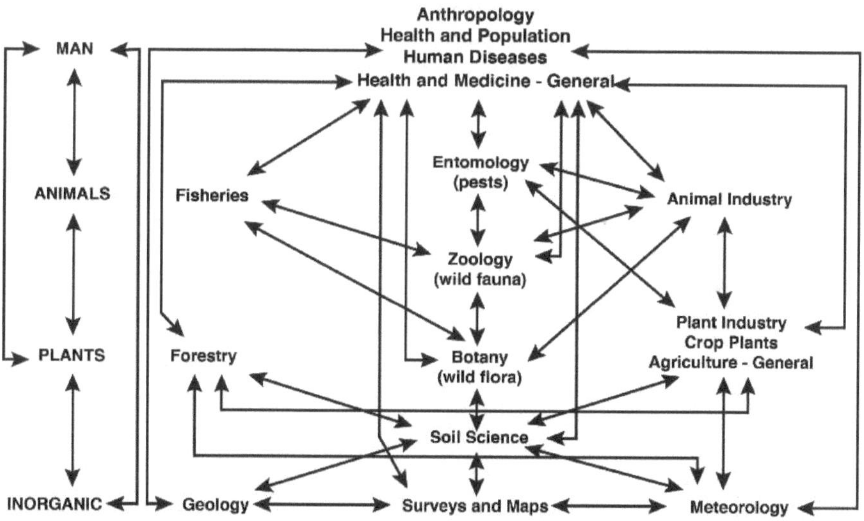

Figure 1.4. *The Colonial Scientific Network of Environmental Management (Worthington to Ravi Rajan [RAV 06]). A framework that resembles many of those that exist today*

The earliest works describing environmental influences on health come from the French geographers Maximilien Sorre [SOR 43], who was credited with the concept of pathogenic complex and Jacques May [MAY 50, MAY 52], the founder of modern medical geography. Sorre [SOR 43] argued that the emergence of diseases depends on physical, biological and social factors and more specifically on the climate, natural biological environment and anthropo-geographical environment. Thus, the environmental conditions, living conditions of the pathogen and characteristics of individuals influence the appearance of a disease [SOR 43]. May's approach focused on the role of the environment in the formation of human diseases and the importance of geography in mapping pathological trends. The pathological factors of disease can only be fully understood if they are considered within a wider understanding of the geographical environment in which they occur. May provided a theoretical framework for studying the environment and geographical factors (in his words, "geogenetics") of pathogen emergence [BRO 04]. Medical ecology thus became a sub-discipline of medical geography [MAY 58, MEA 00].

Box. 1.4. *The origin of the geography of health (according to Oppong & Harold [OPP 10])*

Logan's writings seem so modern that they could have been written by any of the international organizations in the One Health initiative [ATL 10]. But what happened between the 19th Century and today to make Logan and Worthington's writings so relevant?

2

Biodiversity, Cultural Diversity and Infectious Diseases

2.1. Introduction

The human species is the most parasitized animal species, with more than 1,400 species of infectious parasites and microbes [CLE 01]. Over 60% of these parasitic species are of zoonotic origin [TAY 01], meaning that they originate from other wild or domestic animals.

The origin and diversity of parasitism and human infectious diseases in their ecological, evolutionary, biogeographical and historical dimensions have been studied at great length [MAY 58, COC 63, MCN 76, ASH 98, WOL 07, MOR 14, MOR 15, MOR 17]. This chapter begins by exploring major trends in the distribution of human infectious diseases and its links with animal biodiversity and human cultural diversity. Next, the causes and mechanisms that explain this diversity are sought after in human evolution and history, from the first migrations out of Africa to the emergence of agrarian civilizations and the beginnings of globalization.

2.2. Distribution of infectious diseases: links to biological diversity and cultural diversity

Human pathogens are not distributed randomly on the surface of the planet [DUN 10a]. The diversity of pathogens ranges from high latitudes and temperate regions to low latitudes and intertropical regions [GUE 04, DUN 10a]. This latitudinal gradient of diversity of infectious diseases follows the same trends observed for species biodiversity or genetic diversity [SCH 08, MIR 16]. The species richness of vertebrates is thus higher in intertropical zones than in temperate zones, as is the richness of parasites and human pathogens (Figure 2.1). Thus, there is a positive correlation between the species richness of birds and mammals and infectious diseases in humans. Countries with greater biodiversity also exhibit a wide variety of pathogens that infect humans (Figure 2.2).

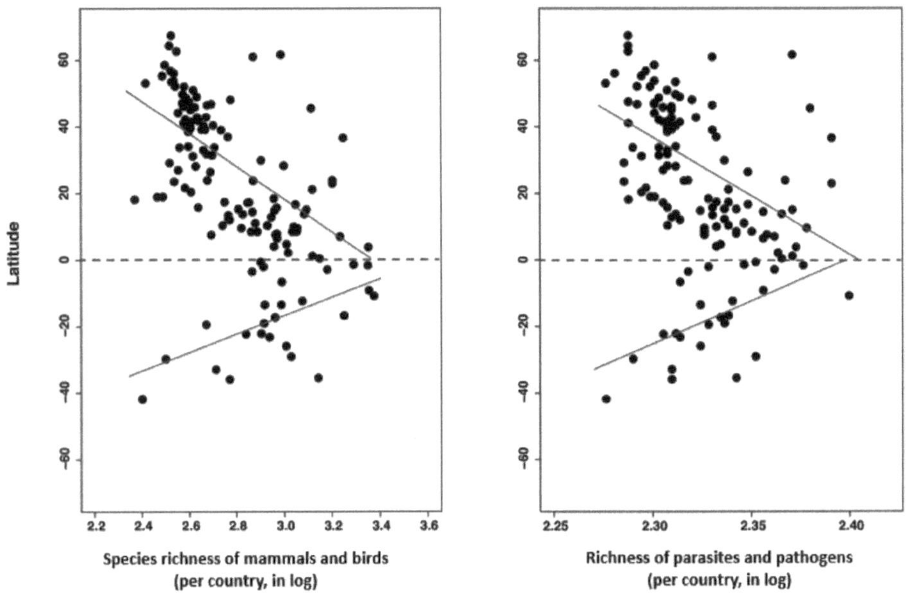

Figure 2.1. *The observed latitudinal gradient (A) for mammal and bird species richness and (B) for the number of parasites and infectious diseases in humans. Data are given per country (in logarithmic scale) according to [MOR 14a]*

Cultural diversity also reflects biodiversity [MAF 05]. Countries with a high cultural diversity estimated through the number of languages spoken are countries with high biodiversity (Figure 2.2). However, countries with a high cultural diversity are also countries that host a wide range of infectious diseases [FIN 08] (Figure 2.2).

Thus, biodiversity, cultural diversity and diversity of human pathogens are closely associated with their geographical distribution because of history and ecology.

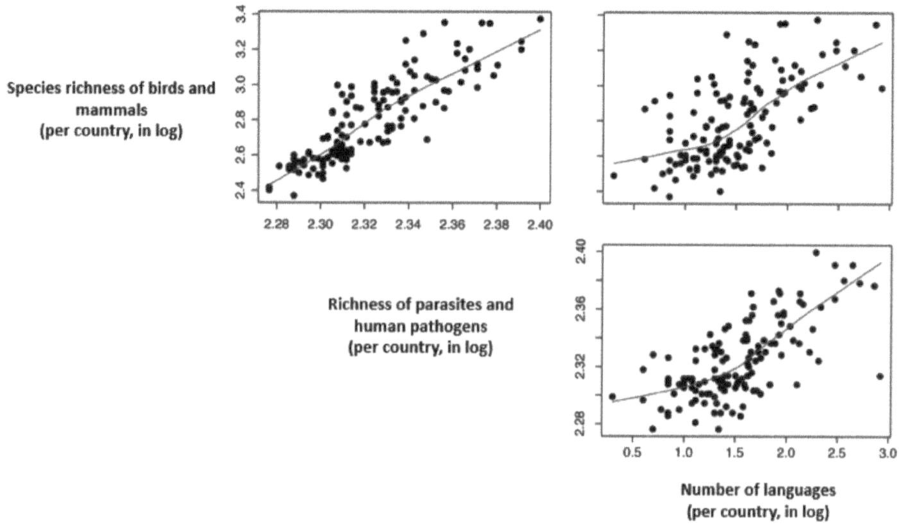

Figure 2.2. *Correlation between biodiversity, diversity of parasites and human pathogens, and cultural diversity (estimated by the number of languages spoken). Data are per country (in logarithmic scale) according to Morand* et al. *[MOR 14a]*

Humans acquire parasites and pathogens in three possible ways: (1) through their descent, meaning that they inherit them from a common primate ancestor; (2) from other wild or domesticated animal species, transmission being direct through the environment or by vectors such as arthropods; (3) from the terrestrial and aquatic environment for many bacteria. Genetic studies of pathogens and parasites make it possible to trace these evolutionary histories.

Three major events in human history provide insights into the changes that affect the wealth and distribution of infectious and parasitic disease communities in humans [WOL 07, MOR 12]: (1) "Out of Africa" dispersal where modern humans brought parasites with them which continued to spread across the planet and for which new pathogens were acquired during the geographical expansion into new biomes of Eurasia, Australia and the Americas; (2) during the emergence of agriculture with increasing sedentarization and the emergence of animal domestication as a source of new parasites and pathogens; (3) during the various stages of globalization with the emergence of agrarian civilizations and the first trade routes to European colonies where many pathogens were able to extend their distribution along land or sea trade routes.

2.3. Origins of parasitic and infectious diseases in non-human primates

African proto-humans and human populations before the African exit were probably predominantly parasitized by species that were common to other non-human primates or were acquired by their offspring. Studies on co-speciation between primates and their parasites have shown that many parasites specific to the human species originate from co-diversification of parasitic lines with primate lines. This includes many organisms such as nematodes, lice, fungal parasites and viruses (see [MOR 12]).

The phenomena of exchanges between more or less related primate species have, however, been frequent, such as the human pubic louse that is more closely related to the gorilla louse than to the chimpanzee louse, which itself is more closely related to the human species than the gorilla [REE 07]. The habitat proximity in addition to geographical proximity contributes to parasitic exchanges between species.

Using a comparative approach, Davies and Pedersen [DAV 08] showed that primate species that are phylogenetically close, and also those living in the same geographic region, share more pathogens than primates who are less closely related or who live in different geographical areas. The sharing of a common habitat or sympatry between related primate species has thus facilitated the exchange and sharing of many species of helminths, protists, bacteria and viruses. This habitat-sharing continues to play a role today in the ecological mechanisms of emergence of new pathogens that originate from primates (as in the case of the Zika virus, see Chapter 3).

Network analyses illustrate the ecological and evolutionary (phylogenetic) components involved in the sharing of parasites by a host community. Gómez *et al.* [GÓM 13a] examined the sharing of parasites and pathogens in primates, using this type of analysis to identify essential species and therefore the species that provide inter-species transmission. The analysis confirms that geographic distribution and niche sharing, as well as the population density of the primate species studied, are the key factors in sharing and inter-species transmission of pathogens.

The example of *Plasmodium* protists, which are responsible for the different forms of malaria, is a perfect illustration of this sharing of parasites between humans and non-human primates. The human species is infected with at least five species of Plasmodium: *P. falciparum, P. vivax, P. ovale,*

P. malariae and more recently, *P. knowlesi*. All these species infect the red blood cells of their hosts and are transmitted by anopheline mosquitoes. According to Liu *et al.* [LIU 10], *P. falciparum* originated from gorillas, which assumes that proto-humans acquired this parasite through a temporary (at the very least) use of forests. *Plasmodium ovale* is also derived from African primates like *Plasmodium malariae*, but the latter infected monkeys of the New World after its introduction during European colonization. *Plasmodium vivax* comes from monkeys of the Old World. *P. knowlesi*, which is related to *P. vivax*, is found in macaques in Southeast Asia. *Plasmodium knowlesi* is a new form of malaria that emerged in Southeast Asia and is linked to the deforestation in this region.

2.4. The first epidemiological transition: "Out of Africa" human migration

The modern human, *Homo sapiens*, probably left Africa about 150,000 years ago [HEN 12]. Humans scattered from Africa to the Middle East, then independently to Europe and Asia, with two major waves for Asia [RAS 11]. Dispersions in the Americas occurred about 15,000 years ago by populations from East Asia who crossed the Bering Strait in two major migrations. The Pacific Islands were populated later by people from Taiwan, about 5,500 years ago.

Human-specific parasites, bacteria and viruses have been used as markers to trace human dispersions [FAL 03, WIR 05, ACH 08, DOM 11], such as human polyomavirus JCV, hepatitis G virus, the tuberculosis agent *Mycobacterium tuberculosis*, leprosy *Mycobacterium leprae* and especially the commensal and sometimes pathogenic bacteria, *Helicobacter pylori*. All of these species, and others, are believed to have accompanied humans during their ancient and more recent dispersions, and they help us better understand the evolutionary human history of our species [REI 13].

Phylogenetic analyses of the hepatitis G virus suggest an early African origin of the virus in humans, which then accompanied the dispersal of humans from Africa to Asia and the Americas [MUE 05].

Phylogeographic studies of the *Mycobacterium tuberculosis* complex also illustrate human diversification and dispersal [GAG 07]. The *M. tuberculosis* complex is thought to have originated from a highly diversified proto-species that infected various species of African hominids. The *M.*

tuberculosis complex underwent a considerable reduction in its genetic variability due to the emergence of modern humans from a relatively small population, followed by their dispersal both in and out of Africa [GUT 05, ACH 08]. Similarly, phylogeographic studies support an African origin of *M. leprae*, the leprosy agent, followed by an early dispersal in modern humans [MON 09].

2.5. Genetic diversity and human migration

During their migrations to Eurasia, modern humans encountered other hominids that had previously left Africa: the Neanderthals and the Denisovans. Genomic analyses show that crossbreeding occurred between these different species [ABI 11]. Importantly for human genetic diversity associated with infectious diseases, the genes involved in immunity were the main beneficiaries of these crosses, which would have allowed the species to colonize Eurasia as well as new epidemiological environments in which the Neanderthals and Denisovans co-evolved with their pathogens for many millennia. Thus, allelic variants of the HLA genes found in Neanderthal and Denisovan bone genes were found in modern humans in Europe and Asia, whereas they are absent from current African populations, confirming the "Out of Africa" crossbreeding hypothesis.

The link between genetic diversity of immunity genes and diversity of pathogens in the environment was confirmed by Prugnolle *et al.* [PRU 05]. They showed that high polymorphism in the HLA gene complex is observed in tropical regions where viral diversity is high. Parasites are a factor in the diversification of immunity genes. It can be assumed that a loss of parasites is not without consequence on the functioning of an immune system that was built under the pressure of parasite diversity (see Chapter 5).

2.6. Animal domestication

Archaeological studies suggest the occurrence of a large-scale domestication of animals beginning about 12,000 years ago. The fertile Crescent of Southwest Asia was the first center of emergence of agriculture with domestication of varied species and presence of the first great human sedentarizations. Then, 7,000 years ago, centers in China, sub-Saharan Africa, New Guinea, Central, Northwest and South America followed independently. The main centers for animal domestication were located in tropical Africa, the Middle East, Southwest Asia, Central Asia, South Asia, Southeast Asia and East Asia. A few animal species were domesticated in the African cradle: donkeys in the horn of East Africa, guinea fowl in West Africa and geese in Egypt. Similarly, in West Europe, domestication only occurred with a few species like rabbits in the Iberian Peninsula and pigeons in the Mediterranean area. Dogs were probably domesticated in Asia before the Neolithic Revolution. The New World contributed to the domestication of llamas and guinea pigs in South America and turkey in North America. The Middle East to the Far East of Asia gave us all the other domesticated species like cats, zebus, aurochs, buffalo, yak, cattle, sheep, pig, goats, dromedaries, camels, horses, ducks, pheasants and chicken. Several animal species must be continually tamed from wild stock, such as the vicuna in South America or the elephant in Asia [DRI 09].

Finally, several commensal species have been associated with human populations that were sedentary for millennia such as the domestic mouse, the black rat or the Norway rat, which originated from Asia. The eras of domestication stretch from nearly 30,000 years for the dog to less than 2,000 years for the rabbit. Similarly, for commensal species such as rodents, coexistence with humans range from 12,000 years before our era for the domestic mouse from the beginning of the Neolithic Revolution in the Near East to about 6,000 years for the Norway rat in the first large cities in South China [DRI 09] (Figure 2.3).

The domestication of animals has resulted in significant changes in nutrition and human health. Morand *et al.* [MOR 14b] reexamined McNeil's hypothesis [MCN 76] on the influence of time of domestication on sharing of parasites and pathogens. Thus, the number of pathogens shared between domestic animals and humans has increased over time since domestication. Network analyzes show that the level of centrality of an animal in the host-pathogen network is correlated with the duration of its domestication. Thus, for a longer domestication period of an animal, such as dog or cattle, the animal has a higher level of centrality in the network. In other words, it is infected more by many pathogens, which also infect many other hosts (domestic and human) in the network. Over time, the addition of a new domestic animal is accompanied by an increase in exchange and sharing of parasites with members of the community of domestic animals and humans (Figure 2.4).

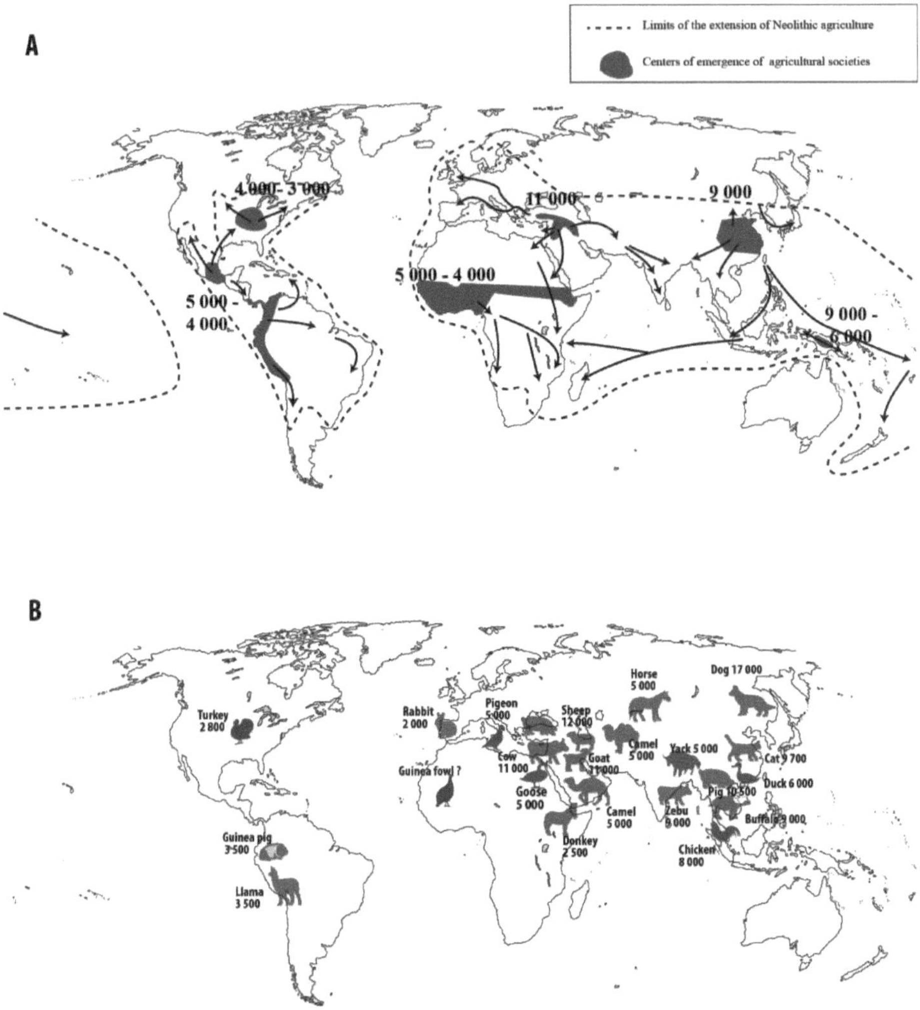

Figure 2.3. *Place and date of domestication of the main animal species (data from Driscoll et al. [DRI 09], figure completed and redesigned by Morand [MOR 16])*

These observations emphasize that in addition to spatial cohabitation, time is an important dimension for the construction of a community of shared zoonotic diseases (as underlined by Wolfe *et al.* [WOL 07]).

Numerous phylogenetic studies have shown that animals were sources of new diseases for humans such as the measles virus from bovines [FUR 10].

Other phylogenetic studies also showed a transfer from humans to domestic animals such as *Mycobacterium bovis* in cattle, which came from a human *Mycobacterium tuberculosis* strain [SMI 09]; or the cestode of the *Taenia* group (the initial parasites of carnivores), which would have infected humans in Africa before they left the continent and before the domestication of cattle and pigs by the neolithic farmers [HOB 01].

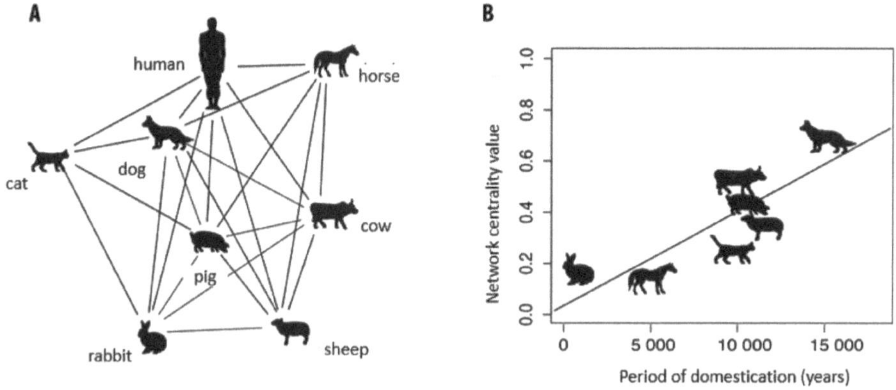

Figure 2.4. *The positive correlation between domestication period and centrality in the sharing of parasites/pathogens in the network of domestic animal associations – in this case, domestic mammals – and humans. The more central an animal is, the more parasites/pathogens it shares with other domestic species and humans. Supplemented by Morand* et al. *[MOR 14]*

The burden of infectious and parasitic diseases likely increased when shifting from hunter-gatherer societies to pastoral societies and sedentary farmers due to an increase in population density and the presence of domestic animals, which are reservoirs of zoonoses. All of these factors favor the diversity and maintenance of infectious cycles [ARA 03]. For example, smallpox, measles and mumps did not affect the first groups of hunter-gatherers, while sedentary farming societies with a high population density became targets for epidemics and the emergence of new pathogens [COC 71, MCN 76]. Finally, opening up new landscapes through deforestation for agricultural purposes promoted the spread of arthropods and transmission of agents such as malaria, yellow fever and dengue [CAR 02], as well as new contacts with telluric bacteria.

2.7. The beginning of globalization

The rise of large-scale commercial networks is associated with the emergence of agrarian societies. They not only exchanged goods but also infectious diseases such as plague, syphilis, smallpox and leprosy. The most famous example is *Yersinia pestis*, the agent of two ancient pandemics: the Justinian plague (541–767) and the Black Death (1346–1353) (Box 2.1). The last plague pandemic was in Hong Kong in 1894, when rats and fleas infected with the bacteria were transported all over the world through bunkers in commercial ships [ACH 08].

> The Black Death struck the Mediterranean area and Europe between 1346–1353. This Eurasian pandemic killed a third of the European population. The writings of many chroniclers from that time have made it possible to accurately retrace the itinerary of the epidemic from its arrival in the port city of Cafta on the shores of the Black Sea during its siege in 1346 by the Tatar troops of the Golden Horde. In less than five years, the entire Mediterranean area and Western Europe was afflicted. The bacillus plague is still present in rodent populations in the Central Asian plains. This bacillus is a soil bacterium that can be transmitted to rodents by fleas. By multiplying and accumulating in the anterior part of the esophagus of the flea, the bacillus prevents the flea from feeding. Another peculiarity is the high lethality of the bacteria for rodents. Thus, epidemic cycles of the plague result in significant mortalities of rodents. Groundhogs in Central Asia and many other species of wild rodents also undergo cycles known as sylvatic plague. Transmission of the bacillus to commensal rats such as the black rat has completely altered the dynamics of infection transmission by humans. The anthropophilic behavior of the black rat exported the disease over large distances through land and sea trade routes that connected Asia to Europe, such as the Silk Road and the Indo-Arab maritime routes. In addition to the role of the commensal rat, climatic variability has played a role in the waves of successive epidemics. Modeling of plague epidemics provides a relatively simple picture of the role of climate variability [SCH 15]. Climate-related variability promoted rodent outbreaks in Asia, followed by transmission of bacillus to commensal rodents and the human emergence of the disease spreading through commercial channels. This spread took about 10 years to reach a Mediterranean port. European dissemination of the disease was ensured in under three years through its connections within the European merchant network. The absence of wild European rodents to ensure a reservoir of the bacillus required reinvasions from Asia.

Box 2.1. *Plague, rodents and climate*

Historian William McNeill [MCN 76] compiled historical references on epidemics that affected China from the Third Century BC until the late 19th Century (Figure 2.5). The dynamics of these epidemics has shed some light on the epidemiological history of civilizations. The first epidemic dates back to 242 BC and it would have struck the entire Chinese empire. The second was only notified in 48 BC. An average of six to seven epidemic events per century was then recorded during the first millennium. After this, there was a succession of large-scale epidemics, both in terms of the number of cities or provinces affected and the number of victims. The true point of inflection in the curve of epidemics was in the 14th Century when 17 events were reported. This reached 42 events in the 16th Century.

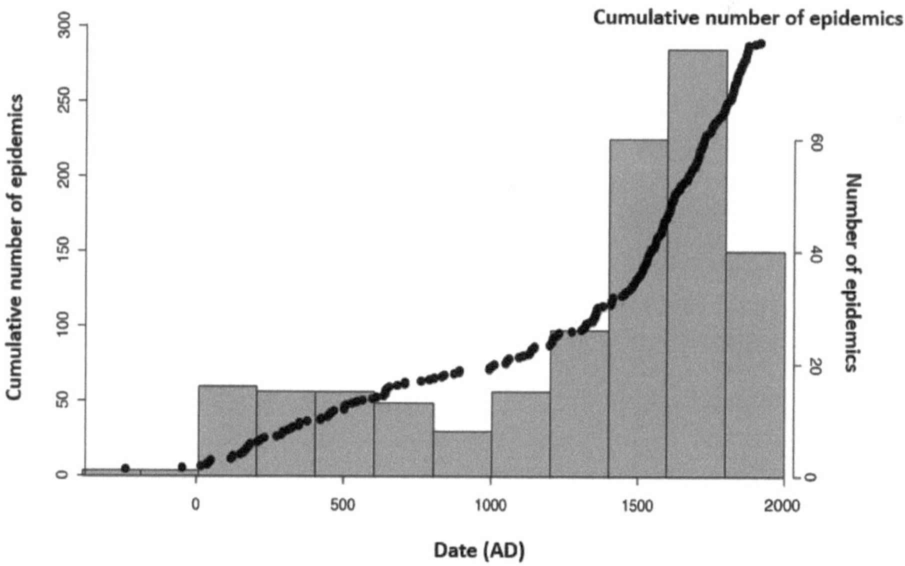

Figure 2.5. *Dynamics of infectious epidemics in China from the Third Century BC to the end of the 19th Century (from [MCN 76])*

The arrival of Christopher Columbus in the West Indies at the end of the 15th Century put an end to nearly 15,000 years of isolation of the Americas. This discovery was followed by the arrival of many conquistadors, with their infectious diseases, which resulted in a demographic devastation of unimaginable proportions [CRO 03]. Demographic statistics established by historians put the total number of people who perished in the first 100 years following first contact with Europeans at 50 million, in other words, 80 to

95% of the Amerindian population. The low genetic diversity of the Amerindian population, particularly with regard to immunity genes, are the main biological mechanisms. The first outbreak was in Hispaniola in 1493 and followed by a succession of epidemics. In 1519, Spanish clerics recorded the loss of a third of the indigenous population on the island.

A genetics study on the mitochondrial DNA of pre-Columbian Amerindian populations, from their arrival in the Americas during the late Pleistocene to the Holocene of agrarian civilizations and current populations shows a significant decline in current genetic diversity. This is due to a collapse in population size subsequent to the widespread mortality of Amerindians [OFA 11].

The large-scale Colombian trading [CRO 03] that came next allowed many crops that were grown in the Americas to spread across the world, such as corn, potatoes and also rubber and cocoa [NUN 10]. But in the case of infectious diseases, the Americas got a lot in return... bubonic plague, chicken pox, cholera, diphtheria, influenza, leprosy, malaria, measles, smallpox, typhoid fever, whooping cough and typhus!

The spread of domestic animals outside of their domestication centers also contributed to the spread of zoonotic diseases [ROS 08]. According to genetic analyses, anthrax, which is from *Bacillus anthracis*, diversified more than 17,000 years ago and then spread across the world with the dispersal of domestic livestock [VAN 07]. Europeans introduced many parasites to the Americas, such as the trichinellosis nematode in America, through infected pigs and/or rats, and the trematode, which is responsible for fasciolosis through infected sheep.

2.8. Conclusion

From the findings presented in this chapter, we can summarize that:

– the diversity of human pathogens is positively correlated with biological diversity suggesting that biological diversity is a source of diversity of parasitic and infectious agents for humans;

– the diversity of pathogens is also correlated with cultural diversity, which is in itself correlated with biological diversity, implying that to address the effects of the biodiversity crisis and its consequences in terms of human health, cultural diversity must also be taken into account;

– human genetic diversity or at least the diversity of immunity genes has evolved in response to parasitic pressure from the environment of human populations. Thus, parasites are a major selective and evolutive pressure force for human populations and their domestic animal populations;

– the diversity of human pathogens has been enriched by animal domestication, sedentarization, population growth and trade during the beginning of globalization. The first historical epidemics that were associated with disease emergence struck Eurasian populations and then human populations that had remained isolated for millennia, such as Amerindian populations.

3

Loss of Biological Diversity and Emergence of Infectious Diseases

3.1. Introduction

Abiotic factors, climatic factors (such as temperature and rainfall) and biotic factors (such as population density and the structure of host communities and reservoirs) are essential variables in the transmission of infectious or parasitic agents [AND 91, MOR 08].

Current global changes (climate change, land use change, biological invasion) are shaking up the epidemiological environment [DAI 96].These changes are responsible for new occurrences and distributions of infectious disease epidemics, as well as the emergence of new infections through the modification of biotic and abiotic factors [WIL 05a, JON 08, DE 08] (see Box 3.1).

In the 1960s, the World Health Organization launched a major program for the global eradication of smallpox, which was successfully completed in 1980. The late 1960s marked the culmination of the triumph of Pasteurian medicine. The use of vaccines, antibiotics and insecticides suggested that all these infections, which had plagued human societies throughout history, could be eradicated. However, subsequent decades were marked by the appearance of new infectious diseases like Legionnaires' disease, Ebola hemorrhagic fever and AIDS. The 1989 Washington Conference on emerging viruses popularized a new term: "emerging diseases". Doctors, microbiologists, virologists and epidemiologists, including many from military laboratories, were invited along by Stephen Morse, a virologist who co-organized the conference [MOR 90a]. Historians such as William McNeill, ecologists such as Thomas Lovejoy (creator of the term "biological diversity") and even theorist Robert May were also present. Morse [MOR 95] defined an emerging infectious disease as an infection that has appeared recently within a population or that existed previously but whose incidence or geographic area is rapidly increasing. The notion of "emerging disease" had already been introduced by René Dubos in his book "Mirage of Health" in 1959. According to Dubos, emergence occurs due to the accumulation of mutations in the infectious microbe, which leads to new co-evolutive ecologies that potentially bring about new health challenges. At a time when public health was trying to eradicate infections and prepare for a new epidemiological transition with the replacement of infectious diseases by modern non-communicable diseases (cancer, diabetes, infarction), Dubos rejected the possibility of eradication with the Darwinian theory of permanent co-evolution between resistance of the infectious agents and the struggle against them. As a precursor, Dubos inserted this co-evolution into the ecological context of a changing planet.

Box 3.1. *The concept of emerging diseases*

The 1989 Washington Conference and the 1991 United States Academy of Sciences "Committee on Emerging Microbial Threats to Health" [LED 92] recognized this coevolving nature as inscribed in the global ecology of emerging diseases and highlighted the difficulty of spatiotemporal prediction of new infectious diseases. Their conclusion was that we must prepare for the unpredictable. The United States' Center for Disease Control and Prevention (CDC) was set up and published the first issue of a new journal "Emerging Infectious Diseases" in 1995, with an article written by Stephen Morse.

In the early 2000s, the Central Intelligence Agency (CIA) published a report on the risks of infectious diseases and bioterrorism to the national security of the United States. This was followed by a Pentagon report that highlighted the need to take climate change and the resurgence of infectious diseases seriously as threats to national security [COO 06a]. The United States believes that preparing for emerging infectious diseases, just as they would for bioterrorism, in situations of maximum uncertainty is "preparedness", in addition to worst-case scenarios developed by successive US administrations [ZYL 13].

Explanations for these new epidemiologies were associated with climate change and its impact on climate variability, increasing global trade, urbanization, overexploitation of biological resources, increasing population pressures and loss of biodiversity [CHI 04, WIL 05a]. The biodiversity crisis is a direct result of altering natural landscapes due to increased urbanization and intensification of agriculture [GIB 10]. These habitat changes appear to be associated with the emergence of new pathogens due to increased contact between wildlife, domestic animals and humans [LLO 09, LIN 15a, HAS 17].

3.2. Epidemiology of infectious diseases

A significant decline in parasitic and infectious diseases has been observed throughout the last century, particularly in developed countries [ARM 99]. This decline started significantly before the extended use of antibiotics or vaccinations came into play. In the United States, mortality from infectious diseases decreased from nearly 800 deaths in 1900 to fewer than 40 deaths per 100,000 people in 1980 (Figure 3.1). It is worth noting the exception of the 1918 influenza pandemic. The mortality rate then rose slightly from 1981 to 1995 to nearly 60 deaths per 100,000 people in the late 1990s. This increase in infectious disease mortality was mainly due to the emergence of AIDS [ARM 99] (Figure 3.1).

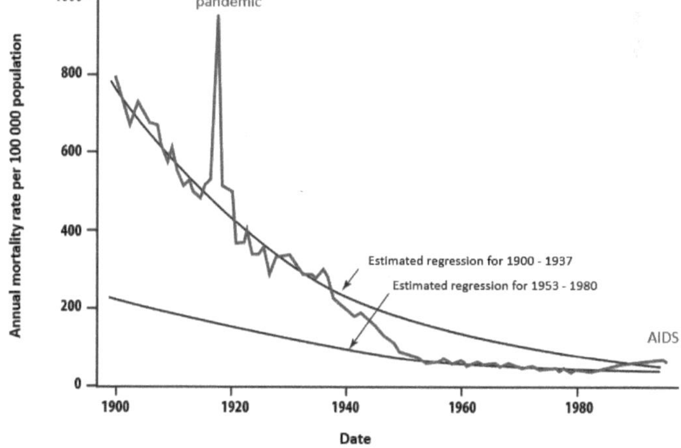

Figure 3.1. *Decline of infectious diseases in the United States during the last century [ARM 99]. Between 1938 and 1952, the mortality rate due to infectious diseases decreased by 8.2% per year, and from 1953 to 1980 the decline was reduced to 2.3% per year. This was followed by an increase of 4.8% from 1980 to the end of the 1990s (taken from [ARM 99]). For a color version of the figure, see www.iste.co.uk/morand/biodiversity.zip*

Similar trends in the decline of infectious diseases have also been seen in developing countries. The economic development associated with the integration of these countries into the world economy has improved the overall level of health of populations. Thus, Martens *et al.* [MAR 10] showed a positive correlation between the degree of integration of a country in the world economy as estimated by the Maastricht Globalization Index and the level of health of the population as estimated by the infant mortality rate. The Maastricht Globalization Index is relatively well correlated with the WHO's Health Development Index and these two indices are negatively associated with the diversity of infectious and parasitic diseases by country (Figure 3.2). The diversity of endemic infectious diseases is reduced as a result of economic growth and integration into the global economy and the associated increase in public health spending.

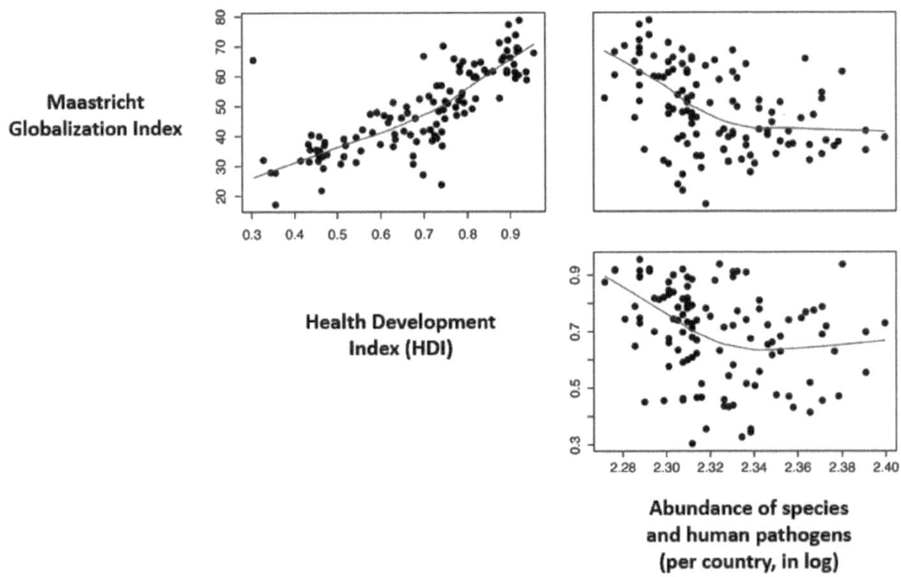

Figure 3.2. *Relationships between the Maastricht Globalization Index, the Health Development Index and the abundance of infectious and parasitic diseases by country. Data on the Maastricht Globalization Index are from Figge & Martens [FIG 14] and data on the Health Development Index and the number of infectious and parasitic diseases are from Morand et al. [MOR 14b]. For a color version of the figure, see www.iste.co.uk/morand/biodiversity.zip*

Recent decades have also seen an increase in the number of newly emerging infectious diseases [JON 08, SMI 14] (Figure 3.3). These emerging infectious diseases are predominantly zoonotic viruses, bacteria and prions (Figure 3.3(a)), in other words, those for which the reservoirs are wild or domestic animals.

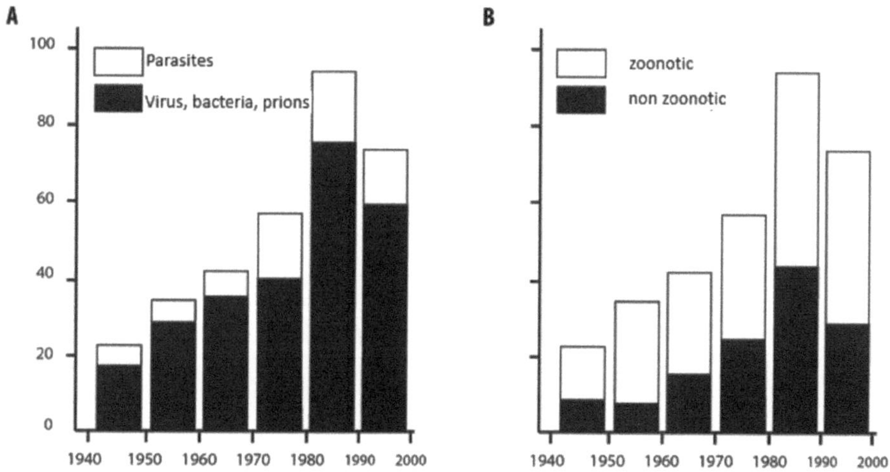

Figure 3.3. *Characteristics of infectious diseases that have emerged in recent decades according to (A) the type of infectious or parasitic agent and (B) the zoonotic or non-zoonotic character (taken from [JON 08])*

The increase in emergence of infectious diseases affects domestic and wild animals and plants (Box 3.2).

Historically, rinderpest epidemics were known in Europe and India and it was only in the 1890s that this disease appeared in the Horn of Africa, subsequent to the introduction of contaminated cattle from India [BLA 00]. It caused 80-90% mortality in wild ungulates in Africa, and the current geographical distribution of many antelope species still reflects the effects of this terrible epidemic [SPI 86, DOB 95]. Rinderpest also caused significant human famine and affected pastoral breeding societies. In his well-known ethnographical study of the Nuers in Sudan, Evans-Pritchard showed that rinderpest profoundly affected the social organization of the Nuers as far as he was able to tell in the 1930s. The Nuers thus shifted from total nutritional and economic dependence on livestock to diversify their activities with the introduction of horticulture.

Marine mammals were not spared either with a significant increase in mortality in various pinniped colonies due to virus outbreaks [FRI 06] (Figure 1).

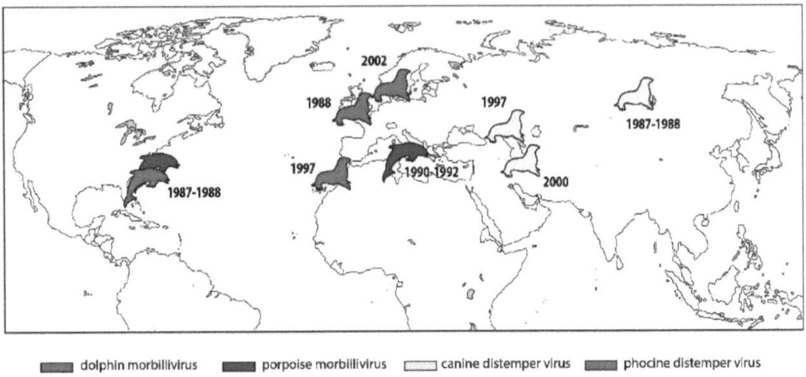

Figure 1. *Examples of events of mass mortality of marine mammals caused by infection from morbilliviruses (according to [FRI 06]). For a color version of the figure, see www.iste.co.uk/morand/biodiversity.zip*

The last decades have shown an increasing number of infectious diseases in fungi-related animals and plants [FIS 12] (Figure 2).

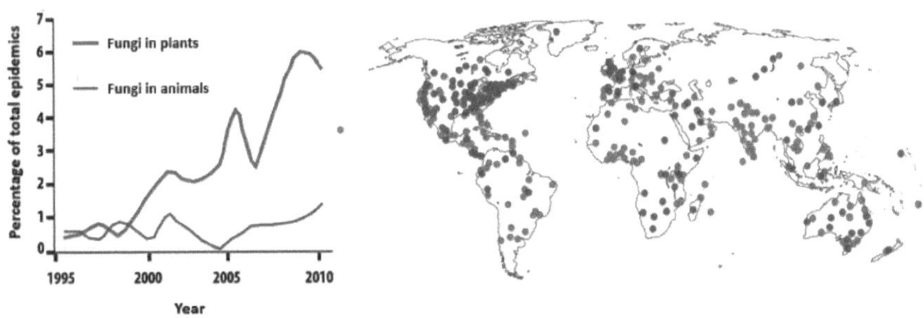

Figure 2. *Increasing epidemics in fungal diseases in plants and animals: on the left, percentage of total number of cases recorded over time; on the right, spatial representation (taken from [FIS 12]). For a color version of the figure, see www.iste.co.uk/morand/biodiversity.zip*

Box 3.2. *Emergences in wild animals*

Some diseases have caused the extinction of certain wildlife species. The *Batrachochytrium dendrobatidis* fungus emerged as a global threat to amphibians in addition to several environmental stressors such as habitat destruction and climate change [FIS 09]. The invasive American frog, *Rana catesbeiana*, is known to spread the fungus to new habitats thus endangering many indigenous amphibian species [HAT 12].

At the same time, the number of infectious disease epidemics has also increased exponentially over the last 60 years [MOR 12a, MOR 14a] (Figure 3.4(A)).

Recent decades have featured an increase in the number of infectious disease epidemics on a global scale, which mainly concern virus-type infectious agents (AIDS virus, Ebola virus, hantavirus, SARS-Coronavirus, etc.) or bacteria (*Legionella pneumophila*, *Bartonella clarridgeiae*, *Borrelia burgdorferi*, etc.) and many of which the origins are to be sought in domestic or wild animals. These epidemics are not accompanied by a significant increase in mortality rates (except AIDS) due to surveillance and public health systems.

An important aspect of these new emergences is their location. The vast majority of these agents were found in developed countries in North America (USA, Canada), Europe, Asia (Japan) and Australia (see map in Jones *et al.* [JON 08]). However, the risks of emergence of these new infectious diseases (especially those related to wildlife) concern countries with high biodiversity and high human density, and thus mainly those located in the intertropical regions [JON 08].

The emergence of new infectious diseases in countries in the intertropical zone followed by their identification (mainly in developed countries) is explained by another aspect of the model: the intensification of human and commercial exchanges.

In recent decades, an increasing number of global outbreaks of infectious diseases with a small and steady increase in new pathogens worldwide has been observed (Figure 3.4(A)) A second trend is the homogenization of the overall distribution of parasites. This homogenization seems to have begun in the 1960s, according to Smith *et al.* [SMI 07], and is characterized by a striking homogenization of global epidemiological patterns. Countries are looking increasingly similar in terms of characteristics of their epidemics in infectious diseases.

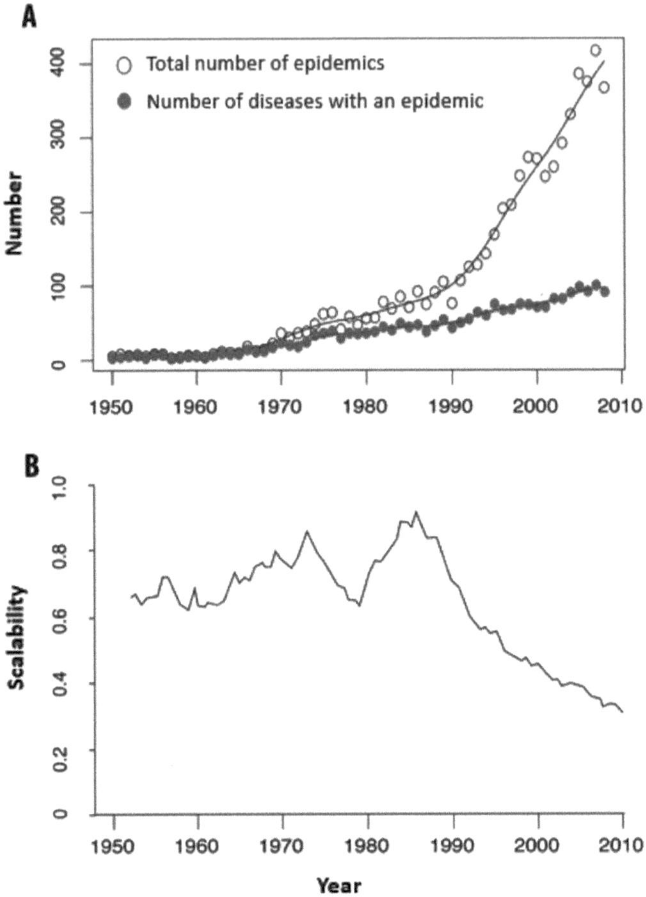

Figure 3.4. *(A) Evolution of the total number of infectious disease epidemics and of the number of different infectious diseases that had at least one epidemic in a given year on a global scale over the last six decades (based on data from [MOR 14b]). (B) Analysis of the modularity of epidemic networks that are shared between countries showing that, per year, the number of countries that share epidemics for the same infectious diseases decreases from the 1980s onwards while the total number of epidemics increases (A). Increasingly, more countries are sharing epidemics (reprinted from [POI 15]). For a color version of the figure, see www.iste.co.uk/morand/biodiversity.zip*

Using network analysis, one can also see a striking decrease in the modularity of country/infectious disease networks over the past decades [POI 15] (Figure 3.4(B)), which highlights that infectious disease epidemics are shared more and more by an increasing number of countries. An epidemic is more likely to reach a large number of countries because of trade links and human displacement.

The observations presented above strongly suggest that global changes affect the worldwide epidemiological environment by promoting the spread of infectious diseases and increasing the risk of pandemics.

3.3. Reservoirs of zoonotic infectious diseases

Woolhouse and Gowtage-Sequeria [WOO 05] characterized the reservoirs of agents that are responsible for new infectious diseases, mainly viruses and bacteria (Figure 3.5).

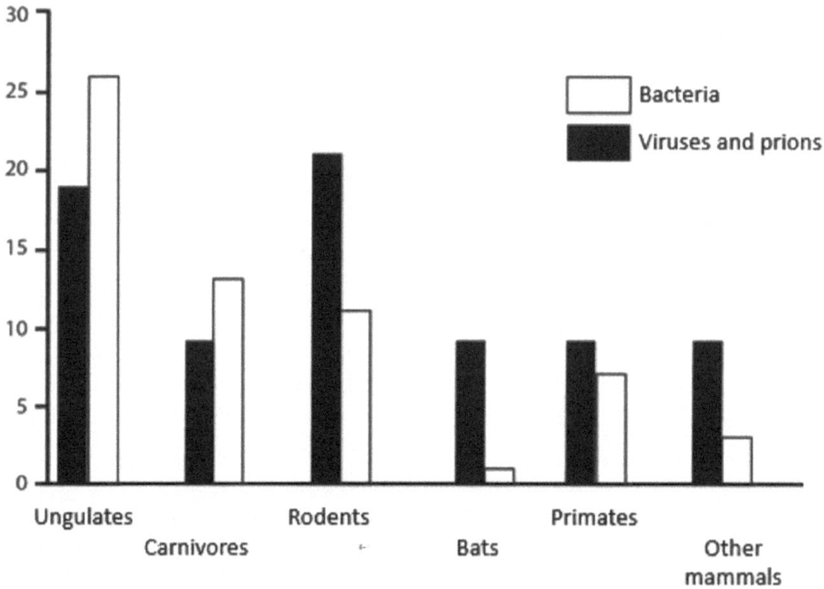

Figure 3.5. *Reservoirs of emerging zoonoses (taken from [WOO 05])*

Ungulates (cattle, horses, goats and sheep) are major reservoirs of new emergences in recent decades, but carnivores (dogs, cats) also play an active role in the spread of emerging infectious agents. Thus, we can appreciate the importance of domestic animals in the diversity of human infectious diseases (Chapter 2).

For wildlife, rodents form the reservoir group that contributes the most to emergence of infectious diseases, followed by primates and then bats.

The incriminated rodents are mostly old commensals of humans (black rats, Norway rats, domestic mice) or "new pets" (prairie dogs, Gambian giant rats). Emerging rodent-related infectious diseases include leptospirosis and hemorrhagic fevers of viral origin (hantavirus, Lassa virus) [MEE 09].

Bats are a source of many emerging viral diseases like Ebola, Hendra, Nipah or SARS-Coronavirus (Box 3.3).

Although the majority of human rabies cases occur from a rabid dog bite, the lyssavirus responsible for this zoonotic disease originates from bats. Carnivores acquired this virus secondarily, which also infected many other animals. Thus, in the early 1900s in Brazil, 4,000 bovines and 1,000 horses and mules died of paralytic rabies. Bats had been seen around these animals trying to bite them and it was the bovines and horses that were actually infected with the rabies virus. This was the first causal link between bats and viral diseases [HAL 07].

Different Ebola viruses have been responsible for several epidemic outbreaks in Central Africa, with the most recent outbreak in West Africa in 2014. The first emergence dates back to 1976 (Ebola virus in Zaire), followed by the Sudan virus, Tai Forest virus and Bundibugyo virus. Transmission of the disease was often due to the consequence of handling bushmeat on markets, as is the case of primates infected in markets in the Democratic Republic of Congo. High mortalities and inter-human transmissions make Ebola a high-risk zoonotic disease. The reservoirs of these Ebola viruses are bats [LEE 15]. Reston virus (RESTV), which is also from the Ebola group, was discovered in macaques in the Hazleton laboratories in the United States in 1989. This virus is non-pathogenic for humans but dangerous for monkeys. It was found in macaques in Southeast Asia and then in bats [JAY 15]. The first infections of the Marburg virus (a German city) involved researchers from a pharmaceutical laboratory who became ill after handling kidney cells from green monkeys imported from Uganda. Epidemics were subsequently recorded in the Republic of Congo in 1998, in East Africa in 2000, in Angola in 2004 and 2005 and in Uganda in 2014. The reservoir was dogfish.

Some emerging viruses belong to the *Paramyxoviridae* family [WAN 08]. The viruses in this family include measles and mumps in humans, and Newcastle disease, rheumatoid arthritis and rinderpest in domestic animals. Four new bat-born paramyxoviruses have emerged since 1994 in Australia, South and Southeast Asia and the Arabian Peninsula. These include the Hendra virus (HeV) in horses and infected humans in Australia in 1996, Nipah virus (NiV) in humans and pigs in Malaysia in 1999, and Menangle virus (MenV) in pigs in Australia in 1997. The different Hendra virus epidemics in Australia have affected horses and some humans who were in direct contact with infected horses. Large frugivorous bats were the reservoirs of this virus. Nipah virus outbreaks occurred in Malaysia in 1998, when farmed pigs and humans became infected. In Singapore, human infestations occurred in slaughterhouse staff where pigs were imported from contaminated areas of Malaysia. Dogfish and small insectivorous bats were reservoirs of NiV. Other Nipah virus epidemics occurred in Bangladesh between 2001 and 2005, and in India in 2001. Infestations were reported to be directly from bats with proven human-to-human transmissions [LUB 09]. The Menangle virus emerged in Australia in 1997 in a large intensive pigsty near Sydney, where two human cases were associated with swine disease. Coronaviridae viruses cause innocuous human diseases for four of these viruses, but two viruses caused two major health crises: the SARS virus (Severe Acute Respiratory Syndrome) with more than 8,000 people infected in some 30 countries and the MERS virus (Middle East Respiratory Syndrome). In 2002, a coronavirus emerged in the province of Guangdong in China, causing the SARS epidemic to be linked to small carnivores like civets, which were sold in bushmeat markets in southern China. The wild reservoirs for this disease were bats [WAN 06]. In the Arabian Peninsula in 2012, the first human case of infection from a new coronavirus that was responsible for a respiratory syndrome was identified. This was the MERS-CoV. Human-to-human transmission was identified with cases being imported into Europe, Asia and the United States. The reservoirs were small insectivorous bats, but human infection was via dromedaries that were infected with the virus [CHU 14].

Box 3.3. *What emerging viruses come from bats?*

What should we remember about the emergence of viruses in bats? First of all, the diseases caused by these viruses have high levels of lethality and are the source of major health crises, such as SARS, Nipah, MERS-CoV and the recent Ebola in West Africa. However, direct viral contamination from a bat to a human is rare. This occurs through the intermediary of primates, carnivores, horses, pigs or dromedaries. These animals are close to us, either phylogenetically like primates or because they have been domesticated for millennia. Finally, two main geographical areas host these emergences: Central and Western Africa and the Asia-Pacific region.

3.4. Emerging infectious diseases and the biodiversity crisis

Anthropogenic changes have been repeatedly challenged as the explanatory factor for the emergence and re-emergence of infectious or parasitic diseases (Table 3.1) with deforestation, land-use change, food production systems and biodiversity change.

The existence of a correlation between biological diversity and the diversity of human pathogens (see Chapter 2) makes it difficult to explain why a loss of biodiversity can increase emerging infectious risks.

Morand *et al.* [MOR 14a] showed that the number of endemic infectious diseases per country in the Asia-Pacific region was highly correlated with the number of mammalian and bird species. They also showed that the total number of zoonotic epidemics in recent decades is positively correlated with the number of endangered mammalian and bird species (according to IUCN criteria), while taking potential biases into account such as population size, economic wealth and health surveillance level of countries. It is interesting to note that the number of endangered human languages in Southeast Asia is also positively correlated with the number of endangered species of birds and mammals [MOR 17a].

Infectious diseases	Mechanisms of emergence	Anthropogenic factors
Malaria	Invasion, expansion of vectors	Deforestation, irrigation
Chagas disease	Expansion of vectors	Deforestation, spread of urbanization
Leishmaniasis	Expansion of vectors, change of host	Habitat alteration
Trypanosomiasis	Expansion of vectors	Deforestation
Cryptosporidiosis	Expansion of oocyte contamination	Poor management of watersheds and wastewater
Cryptosporidiosis	Soil disturbance	Climate variability
Filariasis	Expansion of vectors	Urbanization, irrigation
Bilharzia	Expansion of vectors	Irrigation
Onchocerciasis	Expansion of vectors	Irrigation
Lyme's disease	Loss of predators, expansion of reservoirs	Loss of biodiversity, habitat fragmentation
Cholera	Ocean surface temperature	Climate variability

Leptospirosis	Increased bacterial contact, expansion of reservoirs	Habitat alteration, agricultural development, precarious increase of habitat
Meningitis	Dust storms	Desertification
Dengue	Invasion, expansion of vectors	Urbanization, poor housing conditions
Hantavirus	Abundance of reservoirs, resource variation	Climate variability
Rabies	Change of host	Loss of biodiversity
Rift Valley Fever	Abundance of vectors, heavy rainfall	Climate variability
Encephalitis	Expansion of vectors	Irrigation
SARS-Coronavirus	Transfer of host	Bushmeat, wildlife/domestic animal contact
West Nile Fever	Expansion of vectors	International trade, climate variability
Ebola	Human/wildlife contact	Bushmeat, habitat alteration
Nipah/Hendra	Niche invasion, wildlife/ domestic animal contact	Agricultural intensity
H5N1, avian influenza	Extension of livestock production	Trade, agricultural intensification

Table 3.1. *Infectious diseases: mechanisms of emergence in relation to anthropogenic factors (adapted from [MOR 12])*

If data are taken globally, there is a good correlation among the number of circulating pathogens and parasites, the number of birds and mammals in danger of extinction, and the number of endangered languages on a country scale (taking geographical and demographic size biases of countries into account, as well as economic wealth) (Figure 3.6). Declines in biodiversity and cultural diversity are not independent of the number of infectious diseases.

These correlative and comparative approaches suggest that biodiversity is a source of diversity in pathogens, but endangered biodiversity is a source of epidemics. Biological diversity not only affects the diversity of pathogens, but also potentially their prevalence. Thus, Derne *et al.* [DER 11] showed a negative correlation between the number of mammalian species and annual incidence of human leptospirosis in island states and territories.

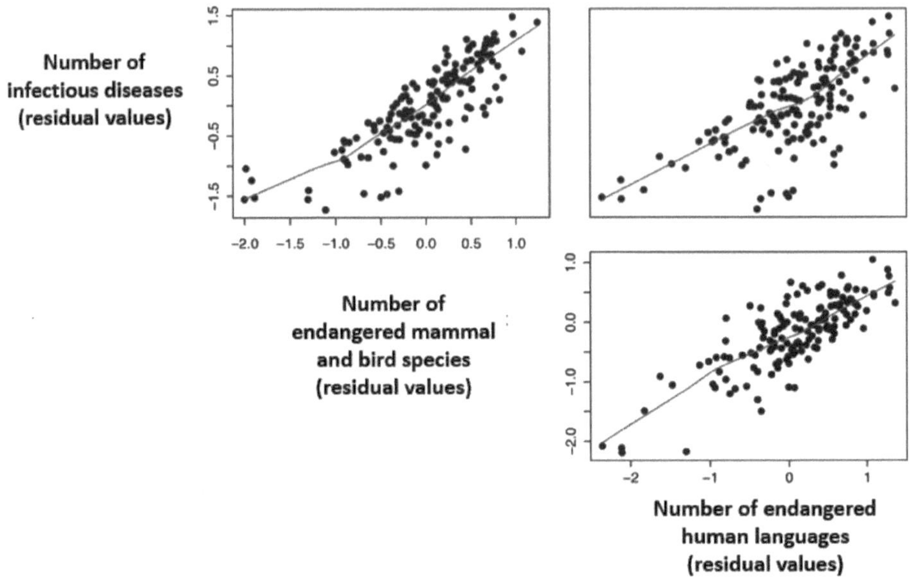

Figure 3.6. *Correlations between the number of pathogens and parasites, the number of endangered mammals and birds (according to IUCN criteria) and the number of endangered languages (according to UNESCO criteria). The variables are corrected for geographical area, demographic size of the country and wealth (GDP per capita) (data supplemented by [MOR 14a]). For a color version of the figure, see www.iste.co.uk/morand/biodiversity.zip*

3.5. Mechanisms of emergence through habitat modification

Loss of habitat and habitat fragmentation usually result in a significant loss of species because the remaining habitats are too small or too isolated to persist or colonize [MOR 10a]. The effect of fragmentation on host-parasite interactions has been the subject of many studies on conservation medicine or emerging diseases [PAT 14a].

No comparative study has yet analyzed the potential links between abundance of zoonotic parasite/pathogen species and habitat changes. On the other hand, prevalence values appear to be affected by changes in fragmented habitats. Thus, Brearley *et al.* [BRE 13] analyzed prevalence changes reported in 19 studies to assess the influence of man-made landscapes. Half of the studies showed an increase in parasite or infectious prevalence with a landscape change. Gottdenker *et al.* [GOT 14] analyzed more studies and showed a similar trend with 56.9% of studies documenting increased pathogen transmission in response to anthropogenic habitat change. Studies on urban landscapes also tend to show an increase in prevalence of infectious agents [HAS 17].

Murray and Daszak [MUR 13] proposed two non-exclusive hypotheses for the emergence of infectious diseases due to habitat alterations. The "disturbance" hypothesis assumes that a change in land use disrupts transmission dynamics in multi-host systems by disrupting inter-species transmission. The "pool of pathogens" hypothesis assumes that land use change exposes new hosts to a greater diversity of pathogens favoring transmission between species.

Long-term studies from observatories on change in land use are particularly useful in understanding the effect of habitat changes on biodiversity. The framework developed by Haddad *et al.* [HAD 15] makes it possible to link consequences of habitat fragmentation to transmission of diseases through these effects on biodiversity (such as reservoirs and vectors) (Figure 3.7).

In their review of observation and experimental devices, Haddad *et al.* [HAD 15] showed that fragmentation reduces species abundance and/or community composition. They also showed that a reduction in size of habitats and increase of borders with other habitat interfaces leads to major changes in the composition of plant and animal communities. Thus, regardless of the aspect of fragmentation (reduction of surface area of different habitats, increased isolation, increase of borders), there will be a change in the diversity and structure of plant and animal communities. The "disturbance" hypothesis can be applied to each aspect of fragmentation and should lead to a reduction in the diversity of pathogenic species and changes in their prevalence. The "pool of pathogens" hypothesis should mainly be applied to a mode of fragmentation that leads to an increase in borders favoring contact between different reservoir and vector communities.

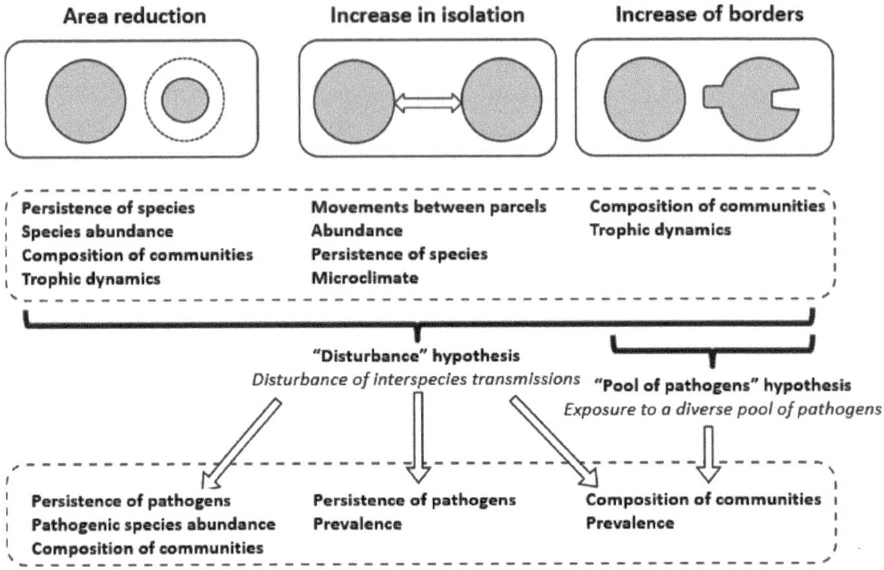

Figure 3.7. *The framework by Haddad et al. [HAD 15] links various aspects of habitat fragmentation to biodiversity. It proposes landscape mechanisms that are responsible for the emergence of infectious diseases according to the hypotheses developed by Murray and Daszak [MUR 13]: the "disturbance" hypothesis and the "pool of pathogens" hypothesis (taken from [MOR 17a])*

3.6. Mechanisms of emergence through community modification

The "dilution effect" hypothesis has been proposed for zoonotic or vector-borne diseases [SCH 01] (taking up the "disease diversity hypothesis" proposed for plants by Elton in 1959). The dilution effect suggests that the abundance and diversity of host species play a protective role in the propagation of pathogens, which is very similar to the "biological resistance hypothesis" formulated for biological invasions.

Species-rich host communities contribute to reducing the transmission of parasites relative to host communities that are not very abundant but have a high proportion of competent hosts. The transmission of a pathogen to an incompetent host is lost, and this type of loss of transmission increases with species-specific abundance. This phenomenon has been observed in several vector-borne diseases such as Lyme's disease [LOG 03] and West Nile

Valley disease [SWA 08], but also in rodent-borne diseases such as the hantavirus [CAR 11].

The first meta-analysis of 13 studies tested the importance of dilution in disease transmission [SAL 13]. It showed that biodiversity had a weak influence suggesting that transmission was dependent on local factors. A second analysis of 90 studies [JOH 15] showed the existence of a dilution effect for various diseases that affect humans, wildlife, livestock or plants. A third study on 61 species of parasites [CIV 15] showed that biodiversity reduces parasite abundance. The dilution effect is robust but its magnitude is related more to the frequency than the population density of the host species.

A change in the predator-prey relationships can affect the transmission of pathogens. Orrock *et al.* [ORR 11] showed that the prevalence of the Sin Nombre virus (SNV) in rodents was higher on islands that host fewer predators of these rodents. A study by Levi *et al.* [LEV 12a] suggested that the increase in recent years of Lyme's disease in the United States is correlated with the decline of the red fox, a key predator of small mammalian reservoirs of the bacteria; this decline can be explained by the competitive expansion of coyote populations.

Ultimately, the consequences of introduction of pathogens, vectors or reservoirs can have dramatic effects on species and local communities [MOR 17]. For example, 10 species of birds among 64 invasive species recorded in Europe have an impact on health [SHI 09]. An invasion may also decrease the transmission of a native parasite. Telfer *et al.* [TEL 05] observed a decrease in the prevalence of two intracellular bacteria, *Bartonella birtlesii* and *Bartonella taylorii*, in indigenous rodent populations, taking the density of another invading rodent into account in their model. This decline in prevalence can be explained by a reduction in transmission of the bacteria by vector fleas.

3.7. Genetic diversity of hosts and transmission of infectious diseases

Host individuals within a given species or population have a genetic variability that is based on heterogeneity in the susceptibility or competence of a given parasite. This intra-species or intra-population genetic diversity must be considered in the understanding of transmission. From a theoretical point of view, genetically diverse populations are susceptible to a wide spectrum of parasites (in terms of number of species) and therefore prove to be beneficial to parasites by harboring a high level of diversity of parasite and pathogen species. Empirical studies in plants, insects and vertebrates support this hypothesis. Reduced genetic diversity in the host favors parasitic transmission and aggravates the negative impacts of parasites by selecting for an increase in their virulence [KIN 12]. In other words, a high genetic diversity of hosts could slow down the spread of parasites and limit their impact.

3.8. Conclusion

This chapter examined the relationship between biodiversity change and transmission of infectious and parasitic diseases from a global to local scale with a brief overview of the mechanisms (Figure 3.8).

While human mortality from infectious diseases has declined dramatically over the past century, recent decades have seen an increase in epidemics of more infectious agents, mostly from wild or domestic animals. This increase in epidemics is accompanied by a greater extension to regional or even global distributions due to an increasingly connected world [TAT 06]. Epidemics are becoming increasingly globalized and frequent.

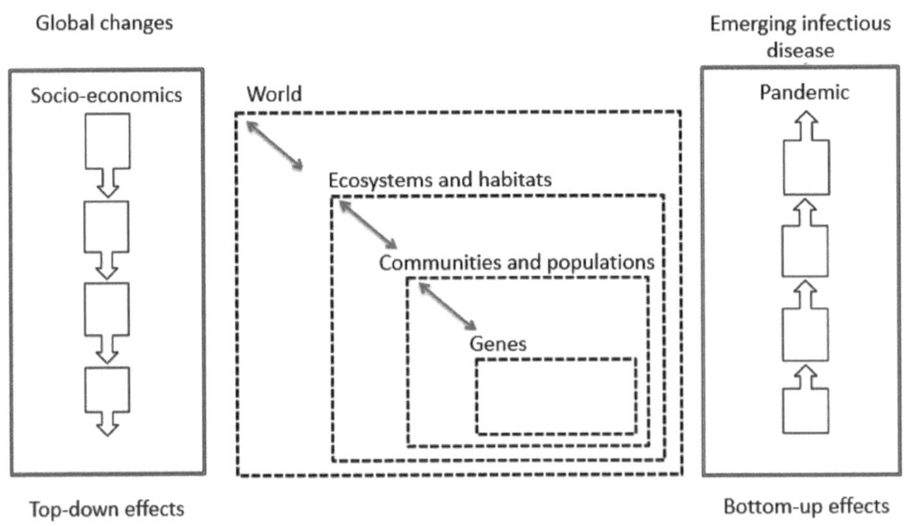

Figure 3.8. *Top-down effects of global changes in biodiversity and bottom-up effects on infectious diseases*

Global and local changes affect all components of biodiversity (ecosystems, habitats, communities, populations, genes) and the interactions between predators and prey, hosts and parasites, as well as human cultural diversity.

Empirical studies and often-correlative analyses show that biodiversity is a source of pathogens, but increases in epidemics and risks of emergence are associated with decreased biodiversity, which is itself associated with land use changes (deforestation, agronomic intensification) and thus favoring interfaces between wildlife, domestic animals and humans [KEE 10].

The decline in biodiversity also concerns parasites and pathogens and this loss of biodiversity does not always have positive consequences on health.

4

Loss of Biodiversity and Emergence of Non-infectious Diseases

4.1. Introduction

The burden of infectious diseases decreased significantly throughout the 19th Century ([ARM 99], Chapter 3). Thus, despite increasing trends in the emergence of new infectious diseases in recent decades, their impact in terms of human mortality or morbidity remains low in Western and developed countries. Infectious diseases, however, continue to affect populations in developing countries.

Developed countries are facing new forms of epidemics and non-infectious or non-communicable diseases such as autoimmune or allergic diseases. This is a new epidemiological transition that has been ongoing for several decades. This transition shows that the rise of non-infectious autoimmune diseases is concomitant with the decline of infectious diseases (Figure 4.1(a) and 4.2(b) [BAC 02]). In parallel, an increase in allergies is associated with a decline in biodiversity as given by the Living Planet Index (Figure 4.1(b) and 4.1(c) [VON 11, HAA 13]).

A link between decreased biodiversity and increased autoimmune and allergic diseases and other so-called diseases of civilization (such as neurodegenerative or cardiac diseases) may be indicative of a single process linked to globalization, with its food and behavioral changes, and global environmental changes. The observed correlation is not evidence of direct causality.

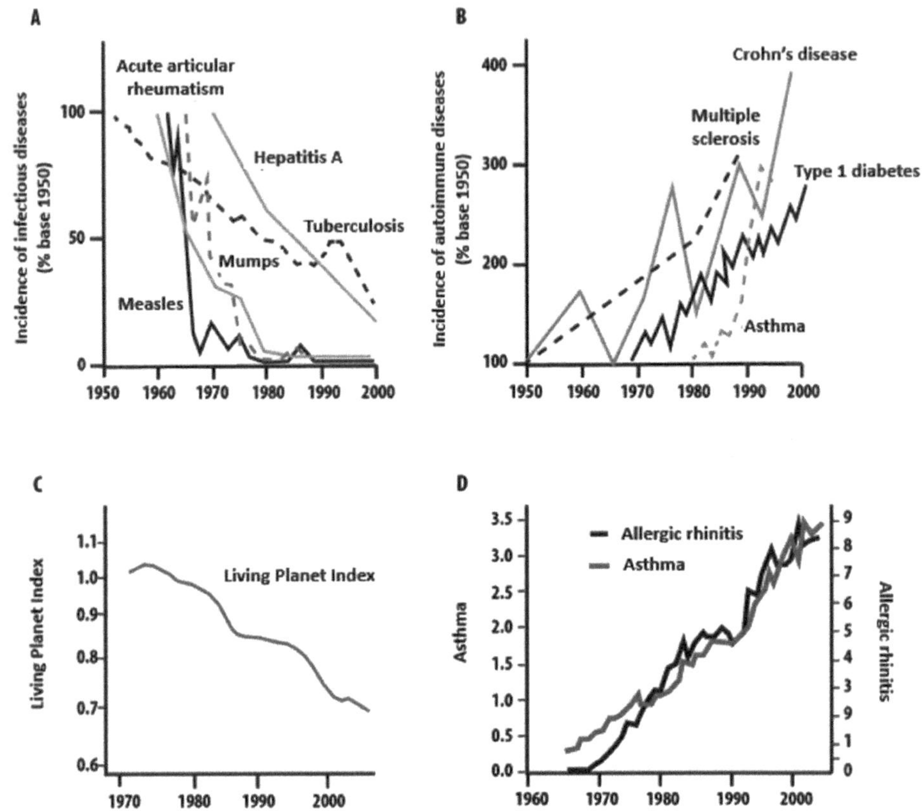

Figure 4.1. *(a) Infectious diseases in the US over time, (b) Different autoimmune diseases in Europe over time, (c) "Living Planet Index" over time, and (d) Asthma and rhinitis allergies in Europe over time (reprinted from [BAC 02, VON 11]). For a color version of the figure, see www.iste.co.uk/morand/biodiversity.zip*

Studies have focused on demonstrating that this link truly reflects the impact of biodiversity changes on biological mechanisms, particularly those resulting from the evolution of fine interactions between biodiversity and humans (as well as between biodiversity and wild or domestic animals). Thus, the "hygiene hypothesis" has allowed us to understand these links between biodiversity and health in a new light. This hypothesis is based on the application of knowledge that results from research on the co-evolution of hosts and their parasites; in particular, the evolutionary consequences of parasite diversity and infectious diseases on the evolution of immune or behavioral defenses of hosts.

4.2. Diversity, host parasite co-evolution and the immune system

As outlined in Chapter 2, communities of human pathogens and parasites develop from a phylogenetic heritage (diversification of primates), from the domestication of animals and all changes in interactions with the environment and other wildlife. Evolution and human history have led human populations to reach densities and geographical distributions that provide exceptional conditions for diversity but also for co-evolution of pathogens. In response to this diversity of pathogens, and temporal and spatial diversification, humans have developed elaborate immune responses [MUE 10]. These responses must also be analyzed in the long term of animal diversification and more particularly within vertebrates for humans.

The vertebrate immune system is classified into two components: innate immunity, which is based on genes that encode receptors to recognize molecular patterns; and adaptive immunity, which is based on receptor rearrangements to identify antigens (molecular elements that are foreign to the body) (Figure 4.2). Innate immunity appeared early on in evolution and is present in all metazoan, vertebrate and invertebrate animals. Adaptive immunity was acquired in vertebrates only with remarkable complexity in jawed vertebrates from teleost fish to mammals and therefore to humans. Adaptive immunity has resulted in the acquisition of immune memory and hence the ability of an organism to respond more effectively to reinfection.

A comparative analysis of immune system genes (Figure 4.2) shows that the evolution of immunity first occurred through the diversification of many genes (often grouped into gene families) during the evolution of multicellular (or metazoan) animals. Let us note that some of these defense genes are derived from bacterial viruses (transposons). Thus, parasite genes have been integrated into the host genome to provide antiparasitic defense. Acquisition of these different types of genes was rarely lost during the diversification of hosts, and we see the complexification of an elaborate defense system against parasitic pressures [COM 99].

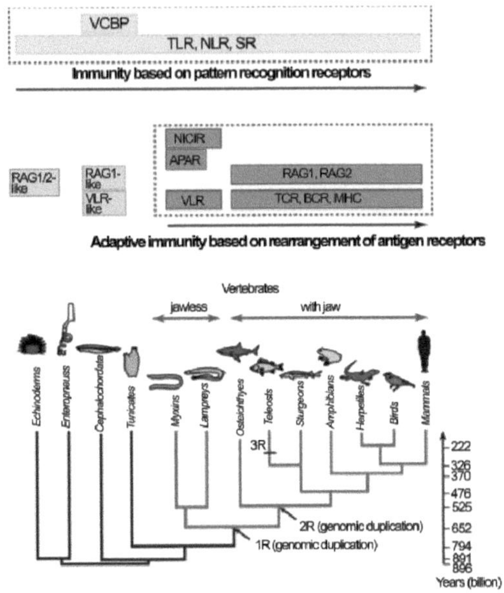

Figure 4.2. *Evolution of immune system genes in vertebrates (jawed and jawless) and in some invertebrate groups. For a color version of the figure, see www.iste.co.uk/morand/biodiversity.zip*

COMMENT ON FIGURE 4.2.– *The immune system is classified into two components: innate immunity and adaptive immunity. Innate immunity is based on genes that encode receptors for pattern recognition. This immunity appeared very early on in evolution and is present in all metazoans, vertebrate and invertebrate animals. These genes include TLR (Toll-like receptor), NLR (Nod-like receptor), SR (scavengers) and VCBP (V-region containing chitin-binding). VCBP is only seen in cephalochordates. Adaptive immunity was acquired in vertebrates, with remarkable complexity in jawed vertebrates (teleost fish, birds, reptiles, amphibians and mammals). Adaptive immunity is based on receptor rearrangements that identify antigens, which are molecular elements that are foreign to the body. Adaptive immunity allows immune memory to be acquired in order to recognize reinfection and organize a defensive response more quickly. Genes that activate recombination, RAG1 and 2 are of viral or bacterial origin (transposons) and are also present in the genomes of sea urchins and cephalochordates. Families of genes particular to vertebrates without jaws (lampreys and myxins) are NICIR (novel immunoreceptor tyrosine-based activation motif-containing immunoglobulin superfamily receptor) or APAR (Agnathan paired receptors resembling antigen receptors). Jawed vertebrates have acquired specific gene families: T cell receptors (TCRs), B cell receptors and the major histocompatibility complex (MHC), which constitute the HLA system. Several genomic duplications were also involved in the diversification of immunity genes, labeled 1R, 2R and 3R (after [FLA 09]).*

Links between the diversity of immunity genes and parasite diversity have been observed in vertebrates, rodents and humans [PRU 05, GOÜ 08]. In Asian rodents, there is a significant correlation between the diversity of the major histocompatibility complex (MHC) alleles and abundance of helminth parasites [PIL 14] (Figure 4.3).

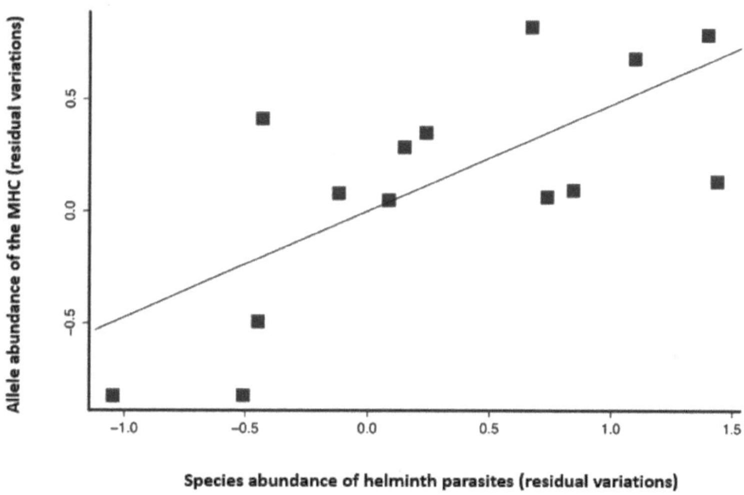

Figure 4.3. *Correlation between allele diversity of the major histocompatibility complex and species abundance of helminth parasites of rodents in Southeast Asia (data from [PIL 14]). Values are corrected for various confounding factors (such as the number of rodents studied or phylogenetic proximity of species)*

The diversity of genes that encode receptors to allow antigenic patterns to be recognized is directly associated with the diversity of parasites and therefore the diversity of exposure [SPU 10].

The advantage of specific adaptive immunity lies in the ability of antigen receptors to diversify genes in order to generate a wide repertoire of cells that are capable of expressing a different antigen receptor – these are the T and B cell lymphocytes.

To give a brief overview, T lymphocytes recognize antigens that have been partially digested by antigen-presenting cells, such as dendritic cells, phagocytic cells and B lymphocytes. The T cell receptors (TCRs) recognize peptide fragments (derived from proteins) or antigens presented by molecules encoded by the class I and class II MHC.

B lymphocytes participate in the humoral component of the adaptive immune system by secreting antibodies. These antibodies recognize particular configurations, or epitopes, in surface proteins and carbohydrates of pathogen microbes. B cells mature in the bone marrow in mammals. These B cells express B-cell receptors (BCRs) on their cell membranes to bind to a specific antigen. The cell then triggers an antibody response [COO 06a].

However, the random nature of the lymphocyte diversification process leads to the generation of certain receptors that recognize self-antigens, in other words potentially harmful T and B lymphocytes that generate autoimmunity. The lymphocytes that carry these receptors are suppressed at their production sites, in the thymus (for T lymphocytes) or in the bone marrow (for B lymphocytes).

The whole process is a total compromise. The immune system needs to generate a diversity of immune cells that is enough to cope with all types of pathogens while avoiding autoimmunity problems, in other words develop immune cells that are harmful to the body. Theoretical models have addressed this problem and suggest that only a significant diversity of the body would be able to optimize this diversification of lymphocytes [DEB 93]. Thus, it would be through diversity of body cells, and their recognition that immunity would be able to recognize non-body cells and therefore pathogens.

Recent studies show that this selection of antigenic recognition diversity is in fact dependent on the diversity of intestinal microbiota – bacteria and protozoa that live in close symbiosis. Microbiota diversity adjusts the diversification of lymphocytic cells and particularly the diversity of the antibody repertoire (immunoglobulins) during the development phase

[GRA 15]. This microbiota diversity is in itself dependent on environmental factors and relies on contact with other microbes, infections or food-related microbes, but also on the impact of antibiotics (see [VOR 14]).

This brief presentation of the immune system provides a few lessons to be learned. The first is that the diversity of immunity genes, at least for the MHC, evolves in response to the species diversity of parasites. The second is that somatic diversity of the adaptive immune response, such as diversity of immunoglobulin production (or antibodies) depends on the diversity of microbiota, both parasitic or symbiotic. The third lesson is that the immune system should be seen as a multi-level network composed of various defense mechanisms (molecular, humoral, cellular), which act autonomously locally (mucus of the skin) or globally (systemic), but communicate between each other and with other physiological functions [DAN 08].

The origin of immune system malfunctions can now be better understood, particularly when the diversity of infectious diseases, microbiota diversity and environmental biological diversity are affected.

4.3. The hygiene hypothesis and the parasitic diversity crisis

The biodiversity crisis also concerns parasite diversity, although empirical data are few and far between and studies are largely theoretical [DUN 09, GÓM 13a, STR 16]. Thus, parasite extinctions are largely inferred under the assumption of co-extinction of specific parasites and their hosts [KOH 04].

On the other hand, there is more convincing evidence for the loss of parasites in humans. This concerns not only major infectious diseases, such as smallpox, but a whole set of parasites that are not very visible or rarely cause pathologies. For example, a 1971–1986 study in a New York hospital showed a 16% annual decline in the number of cases of infection in children in the *Enterobius vermicularis* ward, up until the point of complete disappearance of the parasite [VER 88].

Although helminthiasis or helminth parasites are still common infections in tropical countries, infections from intestinal worms has virtually disappeared in developed countries [LUS 12]. Pest control has been remarkably effective, as it was initially based on hygiene education before the large-scale use of pesticides. In the 1910s, 61% of the inhabitants in Florida were infected with ancylostome-type intestinal worms. These worms can cause severe anemia in the intestine. In the 1930s, they were at 34%. In the 1950s, infection rates had fallen to 18%.

The decline in prevalence of helminth infections in rich countries runs parallel to the increased incidence of immune and inflammatory diseases such as inflammatory bowel ulcers, multiple sclerosis or allergies. Ulcers caused by *Helicobacter pylori* have emerged in many developed countries and now even in developing countries in connection with the disappearance of digestive tract nematodes [WEI 04]. The rise of non-transmissible diseases has suggested that hygiene practices have modified our microbiota (symbiotes and parasites) and consequently led to increased probabilities of allergies and autoimmunity due to dysfunction of immune responses because of an absence of certain parasites [STR 89]. Thus, the prevalence of multiple sclerosis is associated with the prevalence of *Trichuris trichiura* nematode infection [FLE 06].

We must therefore consider the evolutionary role of helminths as regulators of immune response. The eradication of helminths disrupted this regulation causing severe immune diseases. The first theoretical model was based on the premise that sophisticated immune system responses are regulated in order to maintain the individual's homeostasis [WEI 04]. In more simple terms, the immune system of humans responds via two main pathways, TH1 or TH2, depending on the type of infection, bacterial microbes or parasitic worms. In the presence of bacteria, the TH1 pathway is activated, initiating an inflammatory response. When faced with parasitic worms such as nematodes, the TH2 pathway is activated, initiating a cell-type response. The activation of one of these channels, TH1 or TH2, is regulated by the other channel. The pathology of ulcer due to *Helicobacter pylori* is therefore proposed as a result of deregulation of the TH1 pathway which, when unregulated, exacerbates the inflammatory response leading to ulcer formation. The TH1 pathway is no longer regulated by TH2 because there are no more parasitic worms, thanks to hygiene (Figure 4.4).

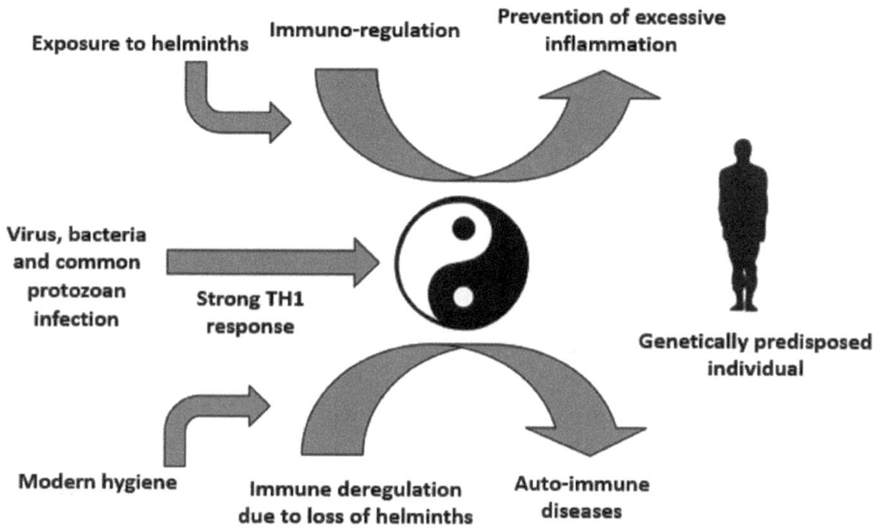

Figure 4.4. *"Helminths and harmony", the hygiene hypothesis model by Weinstock et al. [WEI 04] (adapted from [WEI 04])*

Most studies report statistical associations between the incidence of immune diseases and the prevalence of parasitic infections. However, they are supported by experimental work, which shows that helminths can prevent the emergence of immune dysfunction. Molecular mechanisms have been studied in more detail in laboratory models such as rodents [HEY 14].

These studies have revealed the complexity and multiplicity of immune defenses against helminths and the importance of innate defense immune cells (Figure 4.2) at all stages of parasitic infection by helminths. Cells in the innate immune system, such as macrophages or dendritic cells are the main contributors to certain cytokines (IL-4, IL-5 and IL-13), which promote an anti-helminthic response. Adaptive immunity, which involves cell populations such as regulatory T lymphocytes minimizes the pathology but often does so by blocking the total expulsion of parasites. Thus, for parasitic helminths, what we see is more a tolerance than real resistance. Let us note that B cell lymphocytes can both promote and hinder the immune response. Thus, the induction of immunosuppressive cell populations seems to explain the ability of helminths to attenuate allergies and other immunopathologies [MAI 12].

These studies provide a mechanistic basis for the hygiene hypothesis and epidemiological observations linking decreased prevalence of parasitoses with the appearance of immune diseases [BAC 02].

The ability of helminths to protect against immune diseases has led to their use in therapeutic trials, namely helminthotherapy. Clinical trials have shown that the symptoms of certain immune diseases can actually improve with the administration of helminths because of their immunomodulatory properties [WAM 14, EVA 15, MAI 16, SOR 16]. Research on animal models has shown that diabetic mice infected with different helminths such as *Schistosoma mansoni*, *Trichinella spiralis* and *Heligmosomoides polygyrus* are protected against the risk of diabetes [SAU 07].

4.4. The "farm" hypothesis: biological diversity and allergies

The first indication of hay fever was seen by John Bostock, a doctor who presented his case to the Medical and Chirurgical Society of London in 1819, where he described "a periodic affection of the eyes and the chest", which begins mid-June of each year (quoted by [VEL 13]). Fifty years later, the doctor Charles Blackley, himself suffering from hay fever concluded after inhaling pollen that it was this that caused the disease, and not the sun or the heat. The disease, which was known to affect the nobility began to affect the educated class (the rising bourgeoisie) without affecting farmers, who were in permanent contact with pollen. From these observations, Blackley concluded that the people most exposed to pollen belonged to a certain class (farmers) that was affected the least, and he gave two explanations: education makes one more vulnerable to hay fever or the continued exposure of farmers to pollen protects them. Blackley predicted that if his second hypothesis proved to be accurate, urbanization would considerably increase the prevalence of this condition (quoted by [VEL 13]).

This "farm effect", which closely resembles the hygiene hypothesis has recently been extended to the loss of contact with biodiversity as a source of allergic disease emergence. The hypothesis holds that reducing contact with natural environments and biodiversity (including environmental microbiota) has a negative impact on the diversity of the human commensal microbiota and modifies its immunoregulatory properties.

This hypothesis was put forward to explain differences in the prevalence of allergies observed in Finnish and Russian Karelia, which are socioeconomically distinct regions but have identical geo-climatic environments. Allergy symptoms and diseases (or atopic sensitization) were consistently more frequent among children and adults in Finnish Karelia than among their counterparts in Russian Karelia [LAA 11]. Hay fever was practically non-existent among children in Russian Karelia, whereas it affected nearly 30% of children in Finnish Karelia.

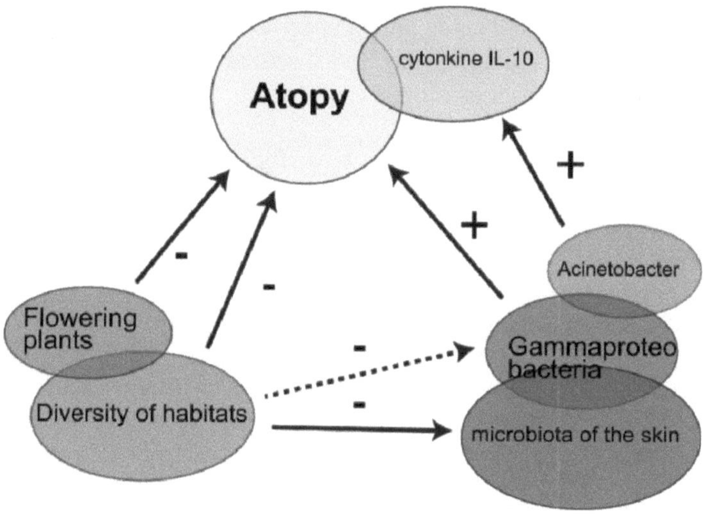

Figure 4.5. *According to the "farm hypothesis", a reduction of contact between people and biodiversity can negatively affect the human microbiota and its immunomodulatory capacity. Hanski et al. [HAN 12] analyzed atopic sensitization (allergic dispositions) in a sample of adolescents living in a heterogeneous region in terms of forest, agricultural and urban habitats. The environmental biodiversity around the dwellings of the adolescents in the study influenced the composition of bacteria in the skin. Compared to healthy subjects, atopic individuals living in a low biodiversity environment harbored significantly lower Gammaproteobacteria diversity on their skin. These have a functional role, through in vitro measurements show the expression of an anti-inflammatory cytokine that is key in immunological tolerance: IL-10. In healthy subjects, the expression of IL-10 is positively correlated with an abundance of the Gammaproteobacterial Acinetobacter genus, which is not seen in atopic subjects (taken from [HAN 12]). For a color version of the figure, see www.iste.co.uk/morand/biodiversity.zip*

Further studies on atopic sensitization in Finland showed that the environmental biodiversity around the studied households, the human microbiota and the presence of allergy were correlated [HAN 12]

(Figure 4.5). People with allergies lived in environments with lower biodiversity as estimated by the amount of woodland and agricultural land than healthy people. In addition, sensitive individuals showed significantly lower Gammaproteobacteria diversity on their skin. The functional link between Gammaproteobacteria and atopic susceptibility was supported by a positive correlation between the expression of the cytokine IL-10 from mononuclear cells and abundance of the *Acinetobacter* genus (a gamma proteobacterium) on the skin of healthy but not atopic subjects [HAN 12].

The Hanski *et al.* [HAN 12] study is an example of a health regulatory ecosystem service. Maintaining diverse landscapes that are rich in various pollen could limit the onset of atopic symptoms.

Recent studies in mice corroborate the immunoregulatory role of commensal cutaneous bacteria in the adjustment of immune and anti-inflammatory responses to environmental allergens [SCH 15a].

4.5. Conclusion: towards an evolving medicine

The results presented in this chapter support the assumptions that a decline in parasite diversity, microbiota diversity and human contact with biological diversity are responsible for the increase in autoimmune diseases (the hygiene hypothesis) and allergic diseases (the farm hypothesis).

These hypotheses are supported by comparative approaches and a detailed study on the mechanisms of immune system responses. The regulation of these responses results from a long co-evolution with the diversity of parasites and pathogenic and symbiotic microbes.

The progress of ongoing studies suggests that personalized medicine is being set up, which takes the genetic background of a person into account, his or her socio-environmental living conditions and the diversity of exposures. Such medicine can be criticized because it only concerns curing or at least attenuating the symptoms and does little for their causes and remedies.

There are evolutionary explanations for diseases, such as autoimmune diseases or allergies, which are consequences of ill-adaptation to current environments. We look at the contradictions and conflicts of our physiological and immune responses due to mechanisms that were put in place by evolution in environmental conditions that have since changed. This evolutionary medicine or Darwinian medicine [NES 96, STE 08] aims to understand the evolutionary and ecological principles at the origin of many diseases in order to improve treatments and prevention.

5

Anthropogenic Stress

5.1. Introduction: a planet dominated by humans and their animals

The increase in human demography is accompanied by parallel increases in overall energy consumption (renewable and non-renewable resources) (Figure 5.1), urbanization and agronomic intensification [EHR 12, SCH 15a]. All these factors have important consequences on biodiversity, animal health, plant health and both directly or indirectly on human health.

By 2014, the United Nations had estimated that more than half of the world's population (54%) would live in cities [UNI 14]. The projections predicted a total urban population of more than 5 billion by 2030. It should be emphasized that urban centers produce around 90% of the world's wealth, accounting for the bulk of energy extraction and living resources used by city dwellers. Urban expansion leads to fragmentation of habitats, loss of biodiversity, changes in local climates and alteration of hydrological systems [SET 10, ELM 13]. Thus, areas of interest for biodiversity conservation are likely to be particularly affected by urban expansion, especially in biodiversity hotspots such as the Mediterranean or Southeast Asia [SET 12a]. Mangroves, dry tropical forests, temperate forests and wet grasslands are subject to the effects of increased urbanization [MCD 13] (Figure 5.2).

Cities require dense road infrastructure to connect with each other as well as with agricultural production or mining environments. Ultimately, cities are the origin of large and varied pollution such as light, sound, sanitary or household waste. These pollutants are the stressors on microbial, plant and animal communities that potentially affect ecosystem health and human health.

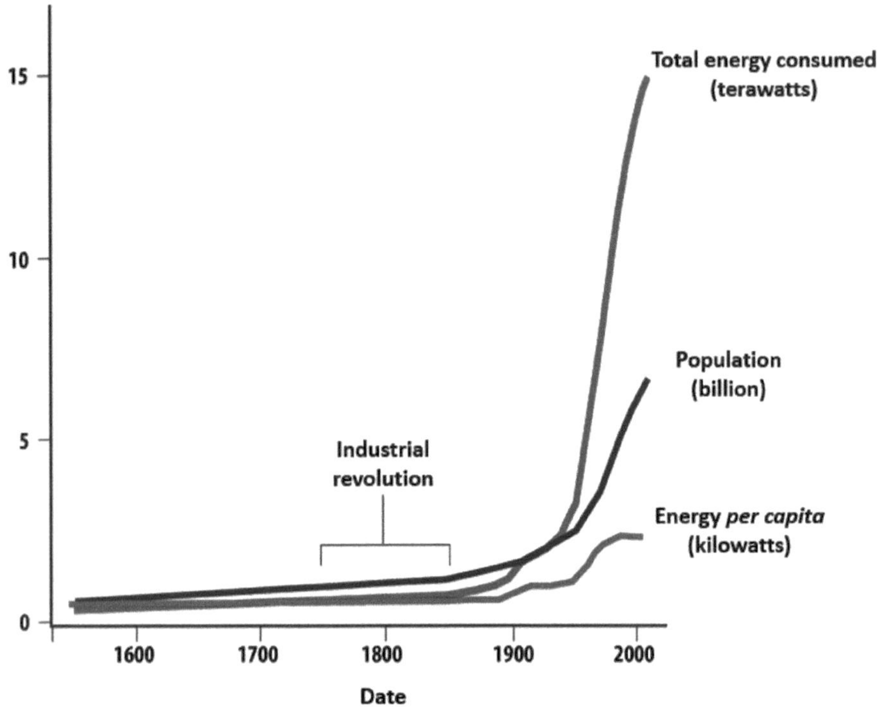

Figure 5.1. *Human population and energy consumption has been increasing since the 17th Century. Let us note the decrease in individual energy consumption in recent decades, which highlights the increase in disparities (taken from [EHR 12]). For a color version of the figure, see www.iste.co.uk/morand/biodiversity.zip*

The extension of urbanization affects agriculture in two ways: on the one hand, it reduces agricultural areas close to towns and on the other, it promotes intensive and industrial agriculture in order to feed the increasing population of urban dwellers. The loss of agricultural land was estimated at nearly 400,000 km^2 between 2000 and 2015 [NEL 10]. Changes in the diets of urban populations to more meat-based and dairy-based diets require the intensification of livestock production with the conversion of agricultural and forestry areas for feeding purposes [SAT 10, RUD 09]. According to a Food and Agricultural Organization (FAO) report in 2006, there has been a direct intensification of land use as pastures as well as indirectly for the production of forage crops. Globally, food crops and cultivated pastures are intensifying in areas with more developed transport infrastructure. The trends of the European Union include declining pasture land and fodder crop production, which are compensated for by importing, such as imports from Brazil's deforested land (Figure 5.3).

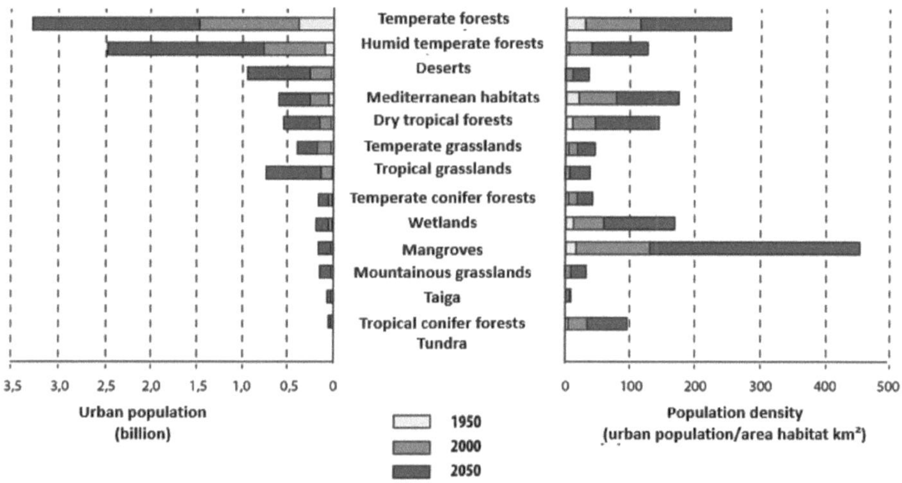

Figure 5.2. *Urban population size (left) and population density per total habitat area (right) by major type of biome (center). The main types of biome are from the World Wildlife Fund (WWF). Urban population data are from the Global Urban/Rural Mapping Program (2000 data). Urban population data for 1950 and 2050 are based on growth rates estimated by the United Nations Population Division (taken from [MCD 13]). For a color version of the figure, see www.iste.co.uk/morand/biodiversity.zip*

This change in land use reflects the increase in total livestock production in terms of cattle (ruminants), pigs, sheep, goats (monogastric) and chicken [THO 10] (Figure 5.4). Planet Earth is thus dominated by vertebrates that have been domesticated thanks to its terrestrial component (Box 5.1). Wildlife, which is diverse in terms of the number of species but reduced for total biomass, only catches a very limited part of the total energy.

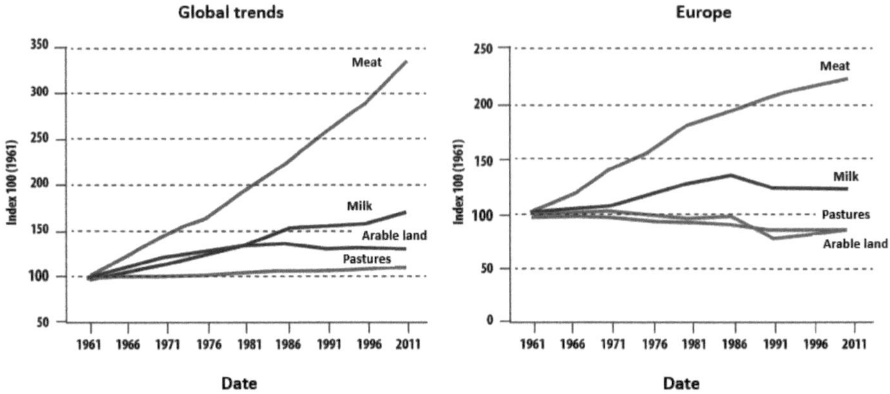

Figure 5.3. *Evolution of land use globally (left) and in the EU (right) for livestock, meat and milk production, and arable land for forage crops (base 100 in 1961) (reprinted according to the FAO [FAO 06]). For a color version of the figure, see www.iste.co.uk/morand/biodiversity.zip*

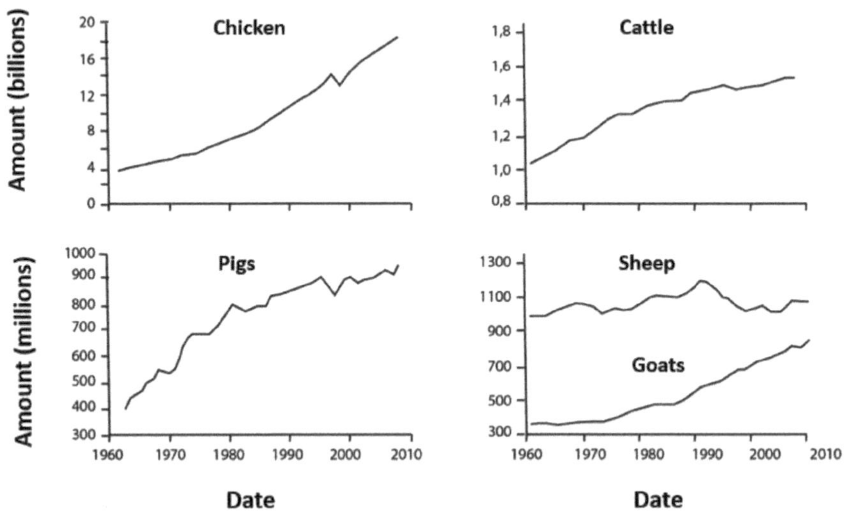

Figure 5.4. *Change in the number of chicken, cattle, pigs, goats and sheep from 1961 to 2008 (taken from [THO 10] based on FAO data)*

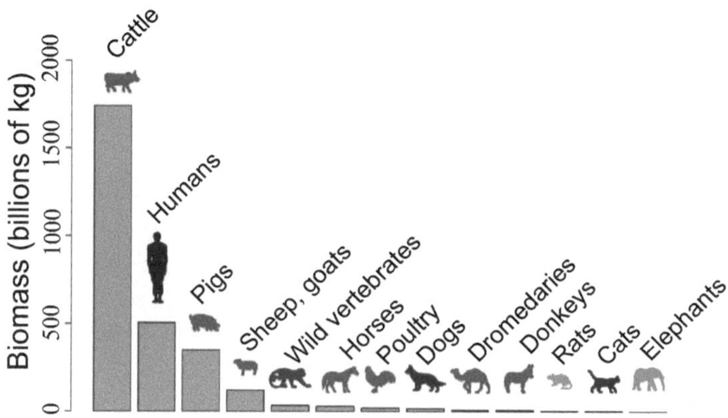

Distribution of biomass of humans and wild and domestic terrestrial vertebrates (based on data from Smil [SMI 02]). Humans account for 18% of the total biomass of terrestrial vertebrate animals, with 61% cattle and 12% pigs. If goats, sheep, horses, dromedaries and poultry are included, domestic animals account for more than 80% of the biomass from terrestrial vertebrates. By adding humans, this leaves a little less than 2% to wild terrestrial vertebrates, not to mention birds and bats. All these domestic animals obviously have food needs and thus large agricultural areas are converted to produce maize or soya.

Box 5.1. *Domestic animals dominate the planet*

The widespread use of inputs, phytosanitary products (pesticides, insecticides), and veterinary drugs (hormones, antiparasitics, antibiotics) are a new source of pollution that affects biodiversity, as well as animal and human health.

This chapter explores the different stressors caused by urbanization and agronomic industrialization and how they interact with the health of individuals as well as animal and human populations.

5.2. Impact of urbanization and road network

Roads fragment landscapes and facilitate human population expansion along with agricultural intensification into new habitats to the detriment of biodiversity. A study by Ibisch *et al.* [IBI 16] showed that vast areas of the planet still lack a major road network and are therefore key areas for biodiversity. About 80% of the Earth's surface is unoccupied. However, far from forming a large homogeneous zone, this surface is heavily fragmented with many zones of varying road densities. Increasing trends and intensification of road networks are a source of threat to animal and plant populations.

First, the road network affects biodiversity through habitat fragmentation, increasing population densities and occurrence of fires [LAU 09]. There are several types of health impacts: direct stress on animal populations and their health by promoting the spread of crop-damaging species (food security risks) and especially vector species or reservoirs of pathogens responsible for infectious diseases. Thus, the spread of dengue virus mosquitoes is associated with the extension of the road network and development of cities along these roads [DUT 98]. The road network directly increases the mortality of wildlife, thus altering species richness and abundance [BAR 10, SUM 11].

An impact of road networks on health is the noise pollution effect. Many studies have shown that chronic anthropogenic noise pollution affects human health [KRY 70, ÖHR 06]. Wildlife is also affected by noise pollution and they appear to react in a similar manner to humans. Aquatic wildlife is also subjected to noise generated by naval traffic at sea or on waterways [AND 11, POP 12].

As pointed out by Francis *et al.* [FRA 12], despite the magnitude of exposure to noise around the world, little is known about how these heavily acoustic environments affect natural populations and communities. Studies have shown that many animals including predators avoid noisy areas because noise impairs their ability to locate prey [SIE 11, MCC 13]. Senzaki *et al.* [SEN 16] showed that the search efficiency of owls decreases with increasing noise from road traffic. The effects of traffic noise on the ability of these birds to detect prey are noticeable up to 120 m from a road. By modifying predator-prey interactions, road noise affects the structure of communities well beyond the alteration of prey-predator interactions; it also changes prey interactions with plant communities in functions such as pollination or predation and dissemination of seeds [FRA 12]. Thus, anthropogenic noise changes the functioning of ecosystems.

Road networks and cities are also sources of nighttime light pollution, which also affects humans and wildlife [NAV 07, SWA 15]. Thus, all the various anthropogenic stressors that affect the ecology and physiology of organisms need to be considered [SWA 15] (Figure 5.5).

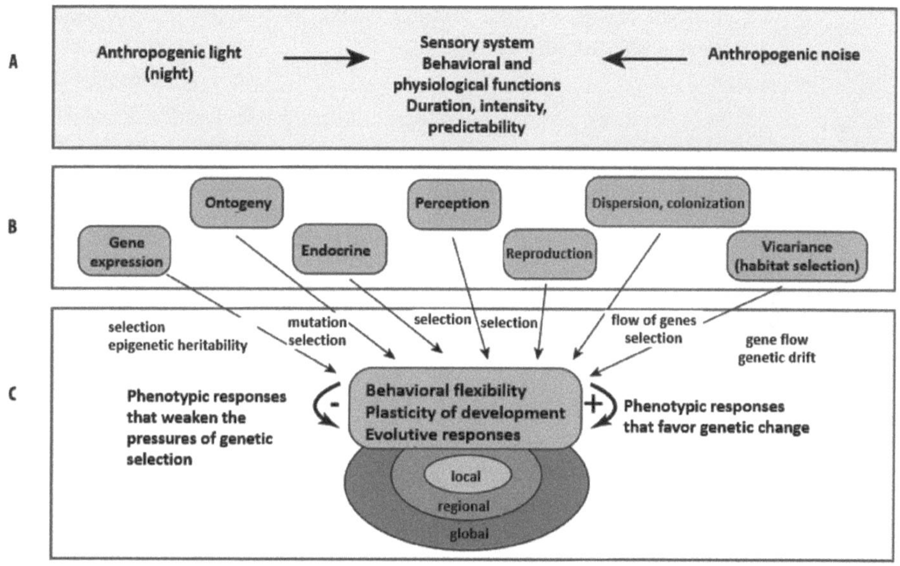

Figure 5.5. *Nocturnal lighting and anthropogenic noise influence the ecology and evolution of organisms. (A) The characteristics of noise and light act independently or synergistically as factors of ecological and evolutionary change as they affect various levels of biological organization (B) through (C) short-term behavioral flexibility, development plasticity and micro-evolving responses (taken from [SWA 15]). For a color version of the figure, see www.iste.co.uk/morand/biodiversity.zip*

The deep physiological mechanisms of responses and adaptations to stressors as well as the consequences of chronic stress on the health of organisms needs to be addressed.

5.3. Physiology of stress and health

Vertebrate responses to environmental stressors are an adaptation to minimize the negative physiological effects of these factors. Natural stressors relate to predation, drought and social hierarchy [ROM 04]; anthropogenic stressors associated with pollution and habitat loss; and ecological stressors, which have a more systemic impact on communities or ecosystems (Table 5.1).

Type	Intrinsic stress factors	Extrinsic stress factors	Ecological driver
Definition	An internal factor or stimulus that results in a significant change in the homeostasis of an animal.	A factor in an animal's external environment that creates stress in the animal.	A biotic or abiotic feature of the environment that affects multiple components of an ecosystem directly and/or indirectly by changing exposure to a suite of extrinsic stressors.
Examples	Pregnancy, breastfeeding, migration, molting, fasting (for example, during the breeding season in some species).	*Anthropogenic*: pollutants, ship strike, entanglement, noise, psychological factors (e.g. perceived threats) *Natural but potentially influenced by human activity*: harmful algal blooms, resource limitation, predator pressure, pathogens, temperature, salinity, naturally occurring chemicals, intra- or inter-specific competition.	Loss of keystone or foundation species, recurring climate patterns such as El Niño, climate change.

Table 5.1. *Types, definitions and examples of stressors (modified from the Committee on the Assessment of the Cumulative Effects of Anthropogenic Stressors on Marine Mammals [COM 17])*

In response to these stressors, the hypothalamic-pituitary-adrenal axis (HPA) triggers a hormone-releasing cascade that stimulates the secretion of glucocorticoids by the adrenal gland [GLA 05] (Figure 5.6). Glucocorticoids help the body prepare for and respond to stress by mobilizing energy reserves and suppressing energy-expensive activities such as suppressing immune responses. The response to stress is an adaptive response. Although high levels of a glucocorticoid like corticosterone are adaptive in the short term, a chronically high level of corticosterone can have a negative impact on survival, reproduction, growth and immune functions [OUY 11].

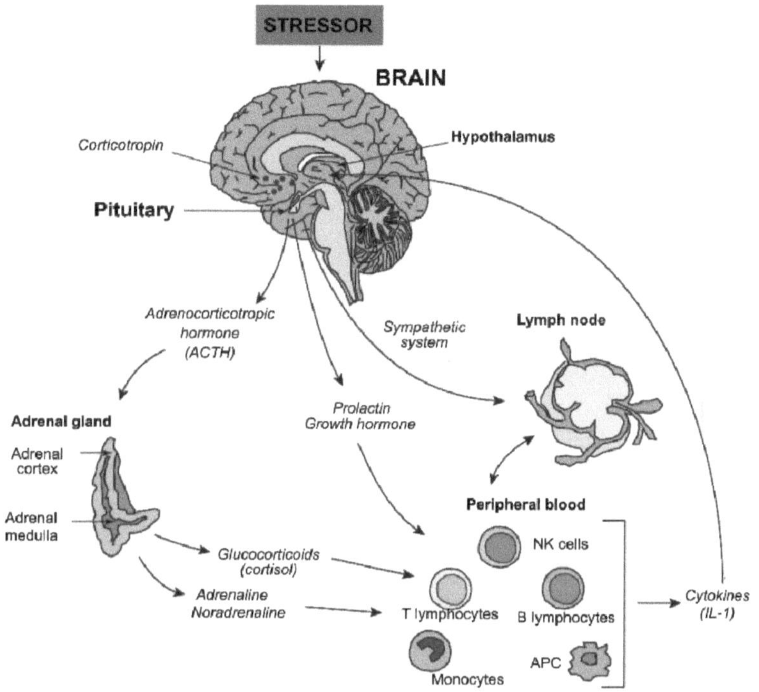

Figure 5.6. *Responses and associated modulations of the hormonal response to stress by the central nervous system. The experience of a stressful situation is perceived by the brain, resulting in stimulation of the hypothalamic-pituitary-adrenal axis (HPA) and the sympathetic-adrenal-medullary axis (SAM). The production of adrenocorticotropic hormone (ACTH) by the pituitary gland leads to the production of glucocorticoid hormones (such as cortisone). The SAM axis can be activated by stimulation of the adrenal medulla to produce catecholamines, adrenaline and noradrenaline, and by innervation of the sympathetic nervous system and lymphoid organs. Immunity is affected by the presence of receptors for stress hormones on the surface of leukocyte cells. Noradrenaline can also modulate immune cell activity. These interactions are bidirectional because cytokines, which are produced by immune cells, can modulate hypothalamic activity (APC: antigen-presenting cell, IL-1: interleukin-1, NK: natural killer) [GLA 05]. For a color version of the figure, see www.iste.co.uk/morand/biodiversity.zip*

The relationships between glucocorticoid levels and parasitic infection have been studied in many animals such as amphibians, birds and non-human primates [HIN 16] with a focus on endangered species that are often subject to significant anthropogenic and ecological stressors [AGU 95, CHA 06]. Thus, in the *Chelonia mydas* turtle, the level of plasma corticosterone is positively correlated with infection from a fibropapilloma virus, which is responsible for an emerging disease [AGU 95]. The physiological levels of stress as estimated by fecal glucocorticoid metabolites, nutrient deficiency and parasitic infection, are also correlated in colobus monkeys (*Piliocolobus badius*) [CHA 06].

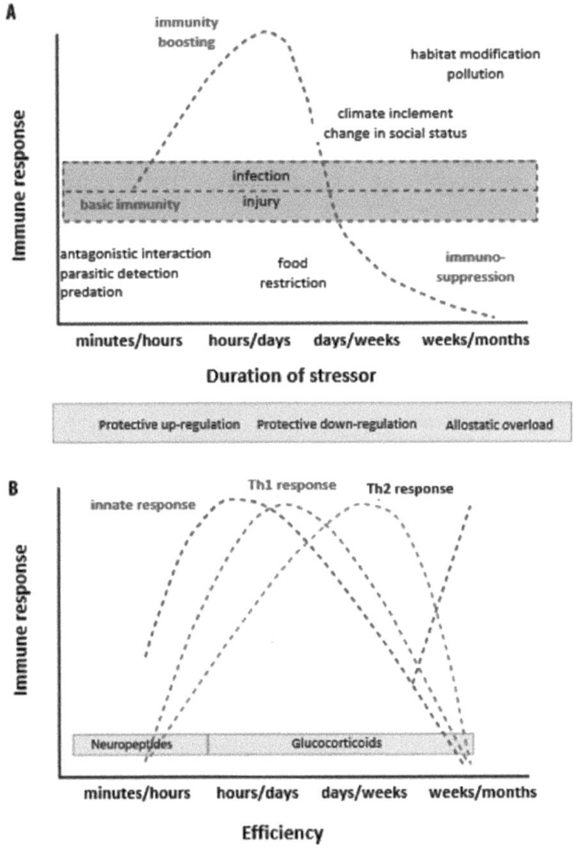

Figure 5.7. *Effects of stressors (habitat, pollution, climate, social status, infection, food restriction, predation, aggression) on (A) immune responses (protective up- and down-regulation, allostatic overload) relative to the time span of the stressors. The shaded area represents the long-term variability of the sensitivity to stressors resulting from individual experience. (B) Changes in innate immune responses, Th1 (pro-inflammatory T cells) and Th2 (antibody response) relative to stressor duration. The repeated increase of innate immunity in response to long-term stressors would be responsible for the emergence of autoimmune diseases. The main mediators of these immune modulations are peptides derived from the central nervous system and glucocorticoids (see Figure 5.6) (from [MAR 05a]). For a color version of the figure, see www.iste.co.uk/morand/biodiversity.zip*

Martin [MAR 09a] explored the causes and modulation mechanisms of stress hormone and immunity. One of the first causes is the redirection of energy resources, such as those required for immunity, to activities that are immediately necessary for survival [SAP 00]. Immune cells and tissues are also actively catabolized to provide proteins and glucose. As the deployment of stress hormones leads to the appearance of self-antigens due to tissue degradation, the immune activity must be reduced to decrease the risk of autoimmunity. According to Martin [MAR 09], immune suppression provides resources over long periods only (days or weeks), because in the short term costs of immunosuppression are likely to outweigh the benefits. In addition, the temporal dimension of stress on innate immune activity and Th1 and Th2 responses must be taken into account (Figure 5.7 and see also Chapter 4).

From this brief presentation of the physiology of stress response, it appears that stressors can have an effect both by increasing susceptibility to infections (and potentially for new emerging infections) and increasing the risk of autoimmune disease.

5.4. Effects of phytosanitation and biocides

Increased agricultural intensification and industrialization of land are accompanied by widespread use of inputs, plant and animal products, and biocides (Box 5.2). These have effects on the various components of biodiversity and the health of wildlife and human health.

According to the European Union (http://ec.europa.eu/health/biocides/policy/), biocidal products are products used to control undesirable organisms that are harmful to human or animal health or that cause damage to human activities with the exception of phytosanitary products. Organisms that are considered harmful include insects, rats or mice, but also microorganisms such as molds or bacteria. Biocidal products are used to control vector-borne diseases (malaria, dengue, chikungunya), food-borne diseases (salmonellosis, listeriosis) and infections acquired in hospitals (multi-antibiotic resistant bacteria). Biocidal products include insecticides, insect repellents, disinfectants, preservatives for materials such as wood, plastics and fibers, or antifouling paints for the protection of ship hulls.

Box 5.2. *Biocides*

The use of phytosanitary agents (pesticides, insecticides, antifungals) is still on the rise, particularly in emerging and developing countries [CAR 06]. The benefits of these agents in terms of food security are far from compensating for their impacts on human health and food health. For example, a recent United Nations report [UNI 17] on the use of pesticides in agriculture and their impact on human rights highlights that "the negative consequences that pesticide practices have had on human health, the environment and society, which are underreported and monitored in the shadow of a prevailing and narrow focus on "food security". The report also recommends that "the environmental and human rights regimes are examined to determine whether the constituent rules are sufficient to protect farm workers, consumers and vulnerable groups, as well as the natural resources that are necessary to support sustainable food systems".

Numerous studies have attempted to estimate the ecological impact of pesticides, their role in the extinction of non-target species such as bees or insect-eating predators and their physiological effects, particularly on the immune system of vertebrates.

Populations of wild and domestic pollinating bees have suffered significant levels of mortality across the globe. Contamination from pesticides remains the primary cause of these mortalities [LHO 14]. The deleterious effects of pyrethroids, deltamethrin and permethrin are well documented in the honey bee. Along with organophosphates (parathion, methyl parathion), these molecules can induce a reduction of the olfactory memory and learning abilities of bees [LHO 14].

Pesticides indirectly affect carnivorous insects and spiders, but decomposers such as centipedes or earthworms are also negatively affected [LEV 12]. The important ecosystem functions and services, such as pollination, crop pest control and recycling of organic matter, are modified by pesticides.

The effects of pesticides can be more chronic as they interact with physiological functions of the body such as the immune system. Pesticides such as atrazine alter the immune response of frogs and their ability to cope with parasitic infections [CHR 03]. Pesticides play a role in the global decline of amphibians in synergy with other anthropogenic factors (habitat loss) and emerging infectious disease [HAY 10].

Aquatic invertebrates may also show increased susceptibility to infections from pesticides. Under experimental conditions, *Crossostrea gigas* oysters subjected to concentrations of pesticides as seen in coastal environments have a higher mortality rate due to infection from a herpes virus [MOR 15].

While pharmaceutical drugs used in both human and animal health have significantly improved life expectancy and food security of the world's inhabitants, their potential effects on the environment and biodiversity have been scrutinized in recent years [ARN 13a]. The effects of drugs on wildlife may be acute with poisonous intoxication leading to a decline in certain species, such as vultures contaminated with diclofenac, which is used in veterinary medicine [CUT 11, CUT 14]. However, the diversity and quantity of pharmaceutical products used leads to a questioning of the effects of genuine medication on the environment [ARN 14] (Figure 5.8).

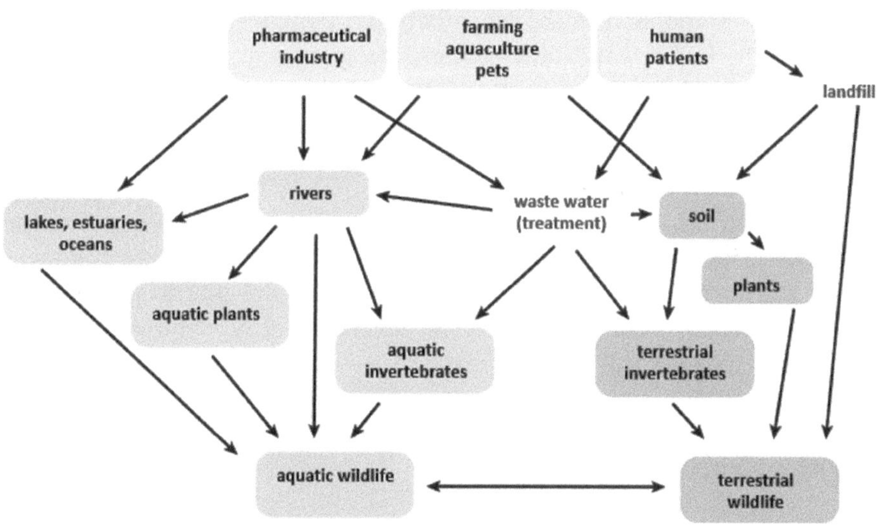

Figure 5.8. *Dispersion of pharmaceutical products throughout the environment. The sources of pharmaceutical products are the industrial manufacturer, livestock, aquaculture, pets and human patients. Pharmaceuticals disperse directly into the environment or through sewage treatment plants. Their dispersal in aquatic and terrestrial environments directly contaminates invertebrate and vertebrate wildlife. Food webs favor their bioaccumulation (adapted from [ARN 14]). For a color version of the figure, see www.iste.co.uk/morand/biodiversity.zip*

Over 4,000 pharmaceutical products are used worldwide for medical and veterinary care, including growth factors used in animal husbandry [ARN 14]. In 2005 alone, 3,303 tons of paracetamol (or Doliprane®, an analgesic), 716 tons of metformin (or Glucophage®, an antidiabetic agent), 444 tons of troxerutin (or Ginkor Fort®), 373 tons of diosmin (Diovenor®) (two vasculoprotectors), 396 tons of acetylsalicylic acid or Aspirin (a nonsteroidal anti-inflammatory drug) and 33 tons of amoxicillin (or Clamoxyl®, an antibiotic) were sold, *inter alia* many other drugs [RAG 14].

When drugs are consumed (and they are not always, as highlighted by Ragon [RAG 14]), the active compounds and their secondary metabolites are found in waste water systems or are released directly into the environment (Figure 5.8). Depending on their physicochemical properties, these compounds may undergo degradation to a greater or lesser extent before they are found in the aquatic (rivers, lakes, estuaries) or terrestrial environments (soil). These products can be absorbed by plants and may potentially contaminate herbivores and thus accumulate in terrestrial and aquatic food webs [SHO 14]. As Arnold *et al.* [ARN 14] highlighted, all wildlife and other ecosystems are likely to be exposed directly or indirectly to pharmaceuticals.

Wildlife is thus exposed to pharmaceuticals and their metabolites throughout their lives and at all stages of development. A study conducted on wild starlings (*Sturnus vulgaris*) showed that the administration of a common antidepressant, fluoxetine, at levels seen in nature affects the behavior and physiology of the birds [BEA 14]. However, organisms are rarely exposed to a single compound. They are exposed to a cocktail of pharmaceuticals and other phytosanitary contaminants which may have synergistic effects on physiology.

5.5. Endocrine disruptors

Descriptions of the effects of endocrine disruptors were initiated by documenting the actions of estrogens, androgens, antiandrogens and antithyroid drugs on wildlife. These effects of response to exposure to ecologically relevant concentrations were measured for many vertebrate species. The founding work on ecology of endocrine disruptors dates back to Colborn [COL 91, COL 95]. The sublethal effects of these products are characterized by dysfunctions in reproduction – effects on reproductive traits of individuals and populations may have ecological and evolutionary implications [SHE 11].

Endocrine disruptors can alter many aspects of the reproductive system: gonad formation, hormone and gamete production, sex determination, egg and embryo formation, and mating signals and behaviors [SHE 11]. Endocrine disruptors are composed of a wide range of products such as herbicides (atrazine), agricultural inputs (nitrates), bisphenol A (present in food plastics), estrogens (17a-ethynyl estradiol), pesticides (DDT, methyl parathion) and heavy metals (mercury, lead) [SHE 11].

Early fetal exposure to endocrine disruptors increases physical and mental health risks. Studies investigating the epigenetic mechanisms involved suggest the existence of transgenerational consequences of exposure to endocrine disruptors [CRE 06], which may lead to genomic changes.

Studies on wildlife have attempted to understand the mechanisms along with the ecological and evolutionary consequences of all phytosanitary or pharmaceutical products. Two new areas of research, immuno-ecology and endocrino-ecology, aim to study the interactions between stressors (intrinsic and extrinsic), physiological responses, immune responses and susceptibility to infectious and non-infectious diseases (Box 5.3). This research is providing new scientific data on the impacts of anthropogenic stressors on individual scales (maintenance of homeostasis, onset of chronic stress), on populations (demography) and communities (trophic and parasitic interactions) [ROH 06]. In this sense, they hold promise for complementary insights *in natura* of traditional ecophysiological tests on laboratory animals.

Measuring stress

Measurements may include glucocorticoids or catecholamines in the blood, urine, feces or hair, pheromones, heat shock proteins and other stress proteins, the relationship between heterophile and lymphocyte immune cells, heart rate, activity rate, and structure of behavioral sequences. The response rates of chemical mediators vary according to the stressors (intensity and chronicity) and their endogenous rhythms.

Measuring immunity

Various approaches are available to evaluate response and immune competence such as measurements of immune reactions after injection of a mitotic agent or an unknown antigen. Functional immune analysis quantifies the proliferation response of lymphocytes and markers can be used to measure individual innate and adaptive immune responses. These include circulating levels of various cytokines or immunoglobulins. Microarrays or RNA sequencing allow direct examination of cytokine expression.

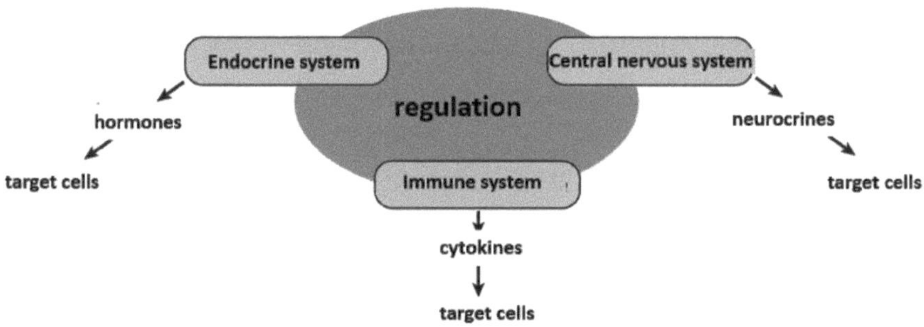

Integrative regulation of the endrocrine, immune and central nervous systems.

Immuno-ecological and endocrino-ecological measurements must be considered within the framework of chemical regulation of the endocrine, immune and central nervous systems (from [NOR 06]).

Box 5.3. *Development of immuno-ecology and endocrino-ecology*

5.6. Antibiotics

The global trend towards intensification of livestock production systems using genetically improved breeds is also accompanied by a growing dependence on antibiotics [SIL 08] (Figure 5.9).

Since they were developed in the 1930s (Figure 5.9) through the work of Alexander Fleming, antibiotics have allowed many infectious diseases of bacterial origin to be controlled and helped reduce human mortality and increase the productivity of breeding. However, evidence on the negative impact of antibiotics on human health, whether used in human or animal medicine, is accumulating [MAR 11, LOO 12, LAN 16].

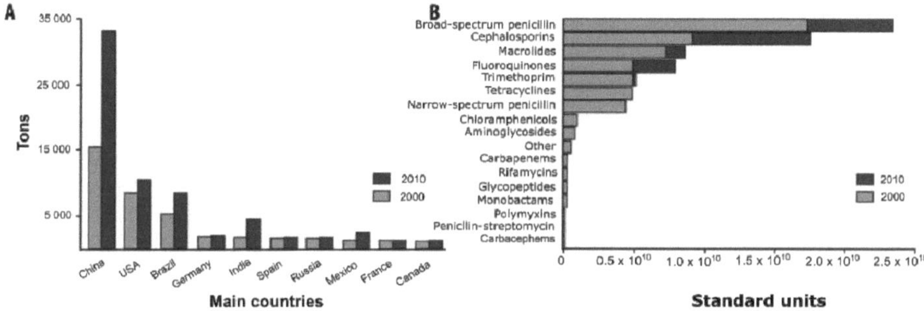

Figure 5.9. *(A) Consumption of antibiotics in livestock production in 2010 and expected trend in 2010 for the ten heaviest consumer countries (from [SIL 08]). (B) Evolution of the consumption of antibiotics in human health between 2000 and 2010 [VAN 14]. For a color version of the figure, see www.iste.co.uk/morand/ biodiversity.zip*

The widespread use of antibiotics over the past 80 years has reduced human mortality. Antibiotics not only attack pathogenic microbes, but also commensal bacteria. Several decades of widespread use of antibiotics has made it possible to understand the consequences on microbial ecology and functional consequences. Thus, antibiotics not only influence the functioning of the immune system and our ability to resist infections, but also our ability to absorb food. Langdon *et al.* [LAN 16] reviewed research on the short-term and long-term consequences of antibiotic use on the human microbiome from early life to adulthood (Figure 5.10). Changes in the human microbiome have implications for diseases such as malnutrition, obesity, diabetes and increased risk of infection by *Clostridium difficile*.

Bacteria are known to respond to stress hormones, including catecholamines, which are secreted during periods of stress. Bacteria use some of these hormones to grow, which can lead to pathogenic processes being induced [FRE 08]). Little is known about the effects of stress hormones on microbiota, which are affected by antibiotics and the onset of bacterial infections.

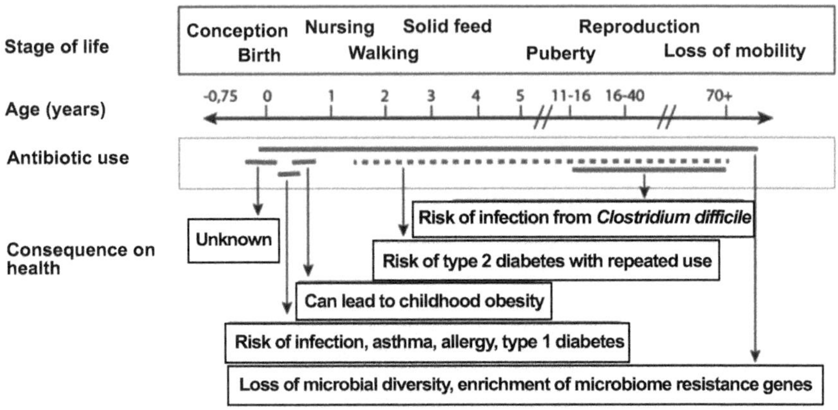

Figure 5.10. *Effects of antibiotics on the human microbiome from early life to adulthood. A single dose of antibiotics can have consequences on longer or shorter timeframes, as shown by the solid red lines. On the other hand, repeated doses of antibiotics over time, shown by the dotted red line, are necessary to see a link, as in the case of type 2 diabetes (from [LAN 16]). For a color version of the figure, see www.iste.co.uk/morand/biodiversity.zip*

5.7. Conclusion

This chapter has shown the consequences of human population growth, urbanization, agricultural intensification and livestock raising as new stressors for animals (and also for humans). These stressors also affect plants and microflora. As such, saprophytic fungi are found to be less capable of decomposing plant litter after exposure to nitrogen-enriched media [VAN 16]. Microbiota also respond to stressors such as heavy metals or antibiotics. This will be further elaborated in the next chapter.

Stress factors directly affect endocrine physiology and immunity, and have a direct impact on animal and human health by promoting susceptibility to infections or by increasing the risk of autoimmune diseases. By modifying social interactions and interspecies interactions, many stressors (noise, phytosanitary products, pharmaceuticals, biocides) have consequences for the ecological functioning and dynamics of populations and communities.

Thus, products used to treat humans, livestock and crops have important consequences on biodiversity, which can in turn negatively affect human health. The ecological consequences of anthropogenic stressors have been explored for several decades through research while their evolutionary consequences are still poorly understood.

6

Biodiversity Response

6.1. Introduction: how life has adapted

Animals, plants, parasites and microbes respond to new environmental conditions on the planet through their physiology and ecology. As they are faced with the pressures of selection imposed by these anthropogenic conditions, new adaptations appear. Some of these adaptations have been around for a long time now, including insect resistance to insecticides, plant resistance to herbicides and bacterial resistance to antibiotics. To quote the title of an article by Palumbi [PAL 01], "Humans as the world's greatest evolutionary force" is explained by the effects of widespread use of biocides and other synthetic molecules to control crop pests and infectious diseases that affect humans and domestic animals. Combating a part of biological diversity and pests for sanitary purposes is accompanied not only by the impact on wildlife and human health (see Chapter 5) but also by the evolution of resistance that leads to the appearance of multiresistant organisms ("super bugs").

It is therefore crucial to understand the responses of biodiversity to new environments (created by intensive agriculture and widespread use of pharmaceuticals) in order to establish new public health guidelines and veterinary and phytosanitary practices to retain their effectiveness and sustainability. This chapter explores the responses of biodiversity to certain health management practices. Evolutionary theories help in understanding the emergence of resistance and also in exploring the evolutionary consequences of new biomedical and biocontrol technologies against infectious agents and agricultural production pests.

6.2. Anthropization and synanthropy

With sedentarization and the emergence of agriculture along with increase in livestock production, human societies have faced an increase in nuisances from many commensal species. Over 3,000 animal species related to human settlements are listed as living which are invasive or potentially damaging to these settlements as well as to food reserves [BON 08, JON 15]. The synanthropic character of these species is their ability to take advantage of environments created by human activities, such as houses, which are extensions of their initial environmental niches. The example of synanthropic rats is illustrative with the Norway or Norwegian rat, *Rattus norvegicus*, and the black rat, *Rattus rattus*. These two species originate from Asia and initiated their association with humans 12,000 years ago for *R. norvegicus* and 6,000 years ago for *R. rattus*. The Norway rat is an infectious species on the banks of rivers in East and Southeast Asia and it invaded the first houses, villages and towns through canals and sewers. The black rat is an arboreal species that builds its nests in the canopy of trees and therefore invaded houses through the roofs. The synanthropy of these rodents is through their role of being major reservoirs for significant diversity of zoonotic pathogens. The story of their long evolution alongside humans demonstrates their adaptive and cognitive abilities to thwart the struggles waged against them, such as mistrust of traps. The adaptive capacities are also demonstrated through their physiology and the appearance of resistance to deratting products such as anticoagulants. The first resistance to first-generation anticoagulant rodenticides (warfarin, chlorophacinone) appeared in the 1950s and signs of new resistance to second-generation rodenticides have also been reported [MEE 14]. This example of rodent resistance to synthetic biocides can be found across all organisms, microbes, parasites, plants, vertebrates and invertebrates, and especially in arthropods.

6.3. Resistance to insecticides

Insecticides are widely used to control insect vectors and insect pests. Resistance to insecticides has been observed in more than 500 species of insects [HEM 00], including over 50 species of *Anopheles* mosquitoes, which are responsible for the transmission of malarial agents to humans

[COR 13]. DDT (dichlorodiphenyltrichloroethane) was first used in 1946 as part of a program to control mosquitoes and eradicate malaria. The first case of resistance to DDT was reported in 1953 in Greece [LIV 53]. The production of other synthetic molecules has also been accompanied by resistance in vector mosquitoes as well as in arthropod pests of crops [WHA 08] (Figure 6.1). Although the cumulative numbers of resistant arthropod species appeared to reach a plateau by the end of 2010, which may have been due to limited investigations of major pests and vectors, the number of resistant cases continues to grow [WHA 08]. Resistance mainly concerns arthropod mites and crop-borne insects, at 63% of cases and 55% of species, compared to arthropod vectors, which are 35% of cases and 37% of species (the rest are mainly parasitoids used in biological control and pollinating insects).

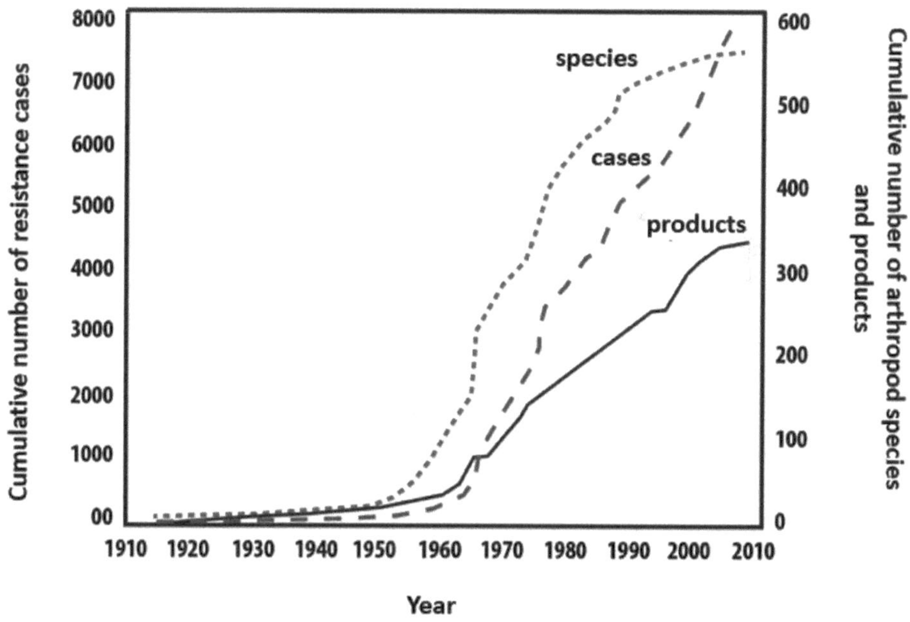

Figure 6.1. *Evolution of arthropod resistance to insecticides: at the end of 2010, there were 7,747 cumulative resistance cases involving a total of 533 resistant arthropod species for a total of 331 products (from [WHA 08])*

The World Health Organization (WHO) defines resistance to insecticides as *"changes in an insect that increase its ability to withstand or overcome the effects of one or more insecticides. When the frequency of resistant insects in a population increases (e.g. through resistance traits being passed on from one generation to another), the efficacy of an insecticidal intervention can be compromised. Increased frequency of resistant insects may be detected through assays that measure insect mortality in response to a particular insecticide, or through genetic tests that detect resistance mechanisms in individual insects"* (http://www.who.int/malaria/media/insecticide_ resistance_management_qa/en/).

The Insecticide Resistance Action Committee (IRAC) (www.irac-online.org) defines resistance as *"a heritable change in the sensitivity of a pest population that is reflected in the repeated failure of a product to achieve the expected level of control when used according to the label recommendation for that pest species"* Biological, biochemical and molecular methods have been used to identify the mechanisms involved in insect resistance to synthetic insecticides. However, this IRAC definition has attracted much criticism in light of new emerging resistance (Box 6.1).

Since the 1950s, there has been an increase in resistance to carbamate, organochlorine, organophosphates and pyrethroid insecticides [COL 17] (Figure 6.2).

The IRAC definition of resistance is problematic when it comes to identification and proactive detection of resistance. When a product results in "repeated failure", it is usually already too late to respond more effectively to resistance. The "expected level of control" is not specific, leaving a broad interpretation of expectations from product users. The definition depends on the "label recommendation for that pest species", and therefore resistance cannot occur in a species that is not mentioned on the label. This excludes taking into account the evolution of resistance in non-pest species. On the other hand, resistance by "heritable change in the sensitivity" of pests is well-defined as the result of evolution, thus providing an objective identification of resistance. It should facilitate proactive detection and apply to resistance in both pests and non-target organisms.

Box 6.1. *Critical appraisal on the definition of resistance given by the IRAC (from [TAB 13])*

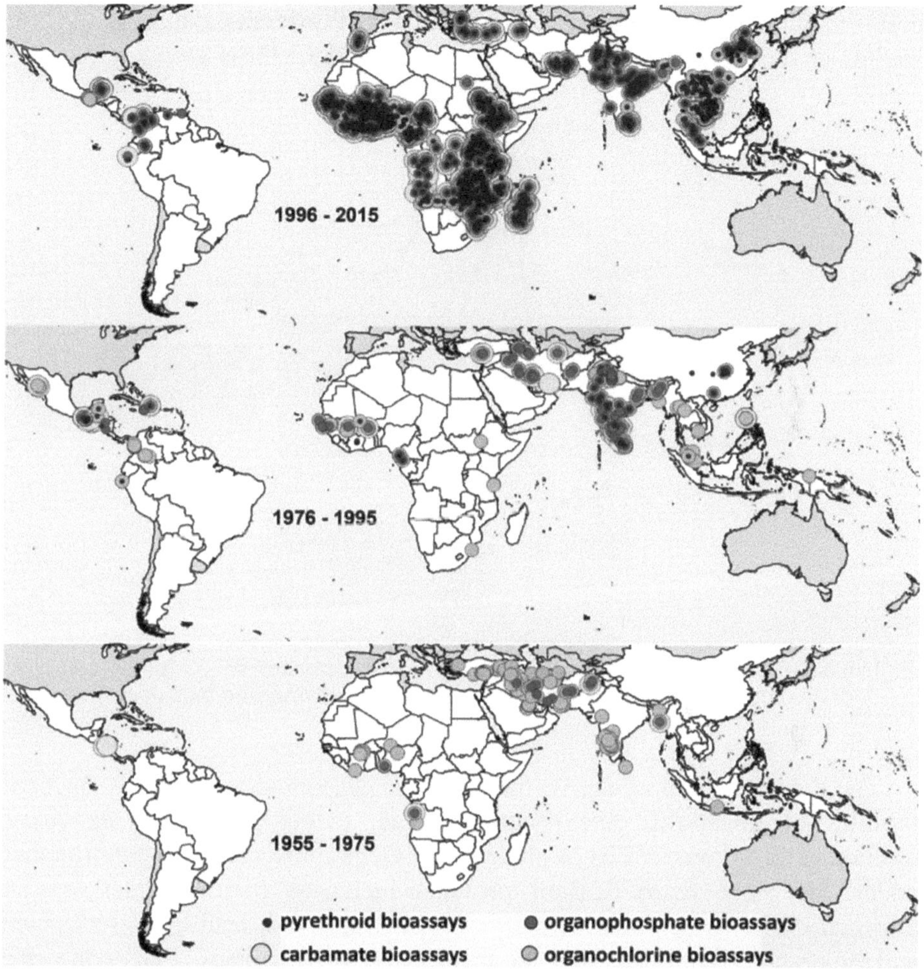

Figure 6.2. *Evolution of resistance to insecticides (carbamates, organochlorines, organophosphates and pyrethroids) of 74 species of anopheles mosquitoes and their species complexes from 1955 to 2015 in 71 malaria endemic countries [COL 17]. For a color version of the figure, see www.iste.co.uk/morand/biodiversity.zip*

In addition to behavioral changes, the main mechanisms responsible for insecticide resistance are: reduction in penetration of insecticides, increase in excretion of insecticide before metabolism, increase in metabolic detoxification of insecticides through overexpression or amplification of certain genes and decreased sensitivity of pesticide target sites. These are point mutations or structural modifications of proteins, such as AChE, sodium channels and GABA receptors, to insecticides [LIU 15] (Figure 6.3).

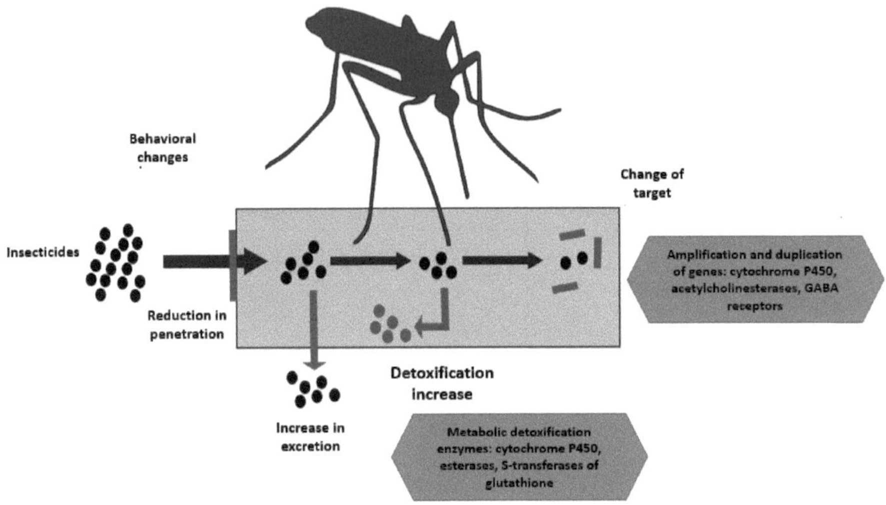

Figure 6.3. *Mechanisms of insect resistance to insecticides (from [COR 13, LIU 15]). For a color version of the figure, see www.iste.co.uk/morand/biodiversity.zip*

The detoxification of active molecules in insecticides involves the main families of metabolic detoxification genes: cytochrome P450 enzymes, esterases and S-transferases of glutathione. Gene transcription regulation that codes for these detoxification proteins increases their production and enzymatic activities. Thus, detoxification is improved, leading to resistance and tolerance. Amplification and duplication of genes that code for cytochrome P450 enzymes are implicated in the development of resistance to insecticides. Pyrethroids and DDT deliver their toxic effects by binding to sodium channels and modifying its properties. Mutations or substitutions in the structural genes of sodium channel proteins in the arthropod nervous system reduce or eliminate the binding affinity of insecticides to these proteins and lead to insensitivity to DDT and pyrethroids [LIU 15]. Mutations of the genes that encode acetylcholinesterases are linked to sensitivity to carbamate insecticides. Finally, changes in the structure of GABA (γ-aminobutyric acid) receptors (the main inhibitory neurotransmitter in insects and mammals and the target site for cyclodiene and phenyl-pyrazole insecticides) have been reported [LIU 15]. Research on insecticide resistance has shifted from the analysis of a single gene involved in metabolic detoxification or modification of target genes, to genomic and metabolic analysis of multi-gene interactions.

The number of malaria cases has halved since the early 2000s [WHO 15a]. This reduction was possible due to the joint use of different prophylactic measures, including anti-vector control using synthetic insecticides. However, increases in mosquito vector resistance for all classes of insecticides is causing us to question the sustainability of synthetic insecticide-based management [RAN 16], as its effect on non-target biological diversity have been notable (see Chapter 5).

6.4. Resistance to genetically modified plants

Genetically modified plants have been presented as a major step forward for phytosanitary defense against rising resistance to synthetic insecticides in arthropod crop pests. The first generation of transgenic cotton that expressed insecticidal proteins from the bacterium *Bacillus thuringiensis* (Bt) occurred in 1996. This was the first Bt cotton to express the protein Cry1Ac, which was developed by the company Monsanto under the trade name INGARD® [AND 08]. This development followed in other commercial plants such as corn.

In 2004, there were no cases of insect resistance to Bt transgenic plants while simultaneous studies focusing on the theoretical aspects of evolutionary genetics were alarmed by the emergence of resistance [BAT 05]. Similarly, little or no effect was seen in non-target species and humans [MEN 03, SAN 11]. The first case of resistance was in 2005. In the early 2010s, nearly 70 million hectares of transgenic plants were planted, and five of the 13 major crop-attacking arthropods had populations that were resistant to Bt transgenic plants [TAB 13] (Figure 6.4).

6.5. Resistance to antiparasitic drugs: the example of artemisinin

The emergence of resistance to synthetic antimalarials was observed for three of the five malaria species that are known to affect humans: *P. falciparum*, *P. vivax* and *P. malariae*. This came with the appearance of cross-resistance wherein resistance to a drug confers resistance to other drugs belonging to the same chemical family or which have similar modes of action (WHO 2014). Thus, over recent decades, antimalarials have had to be withdrawn from the market after the emergence of parasite resistance against them.

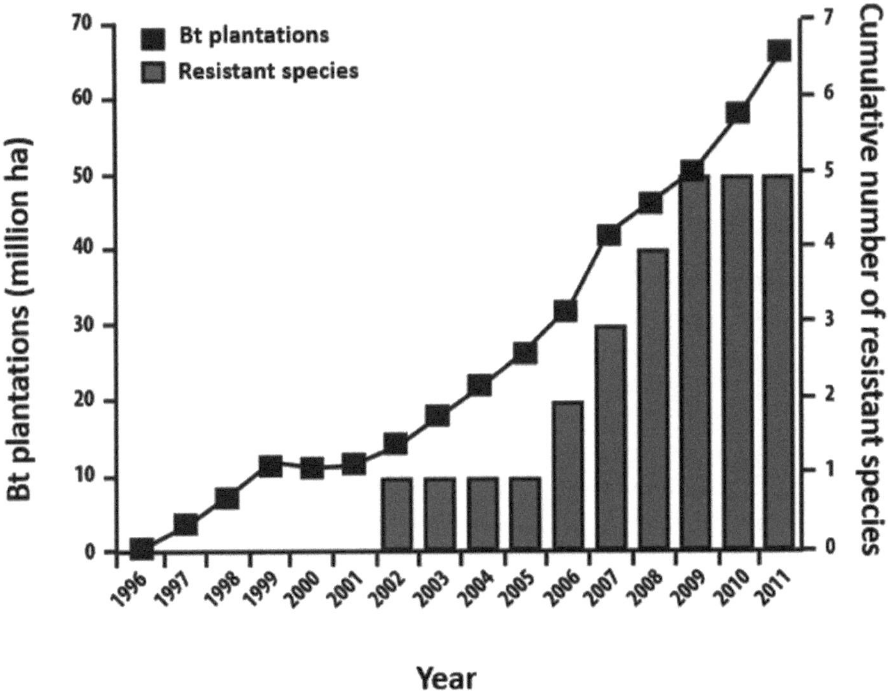

Figure 6.4. *Bt transgenic crop plantation on a global scale and resistance seen in the field, in cumulative numbers of insect species with reduced resistance or reduced efficacy as evaluated in the field. Bt cultivation areas increased from 1.1 million hectares in 1996 to 66 million hectares in 2011 (from [TAB 13]). For a color version of the figure, see www.iste.co.uk/morand/biodiversity.zip*

The effectiveness of many antimalarials is limited by drug resistance and recent evidence suggests that parasites have become resistant to new agents. However, the magnitude of resistance varies; so in many cases, drugs with resistance problems nevertheless offer good efficacy for malaria treatment and control. Resistance has been described to a greater or lesser extent for almost all available drugs [CUI 15]. Resistance to quinine, the oldest antimalarial drug, was first reported in Brazil at the beginning of the 20th Century [CÂN 14] and then in Southeast Asia in the 1980s [PUK 94].

Resistance may be consecutive to mutations for the target genes and conferring resistance, but the mechanisms for many resistances are unknown or little known.

The arrival of artemisinin appeared to mark a favorable turn in the fight against Plasmodia. Artemisinin is a "gift of Chinese medicine" according to Tu [TU 11]. Qinghao, the Chinese name for artemisinin *Artemisia annua L.*, is quoted in the "Handbook of Prescriptions for Emergencies" by physician Ge Hong from the Ming Dynasty (283–343). In order to alleviate the symptoms of malaria, the handbook prescribes "a handful of qinghao immersed with 2 liters of water, wring out the juice and drink it all" (quoted from [TU 11]). It was rediscovered by Tu Youyou in the early 1970s at the height of the Cultural Revolution (1966–1976) and in the greatest secrecy [HSU 06]. A synthetic substance was produced in 1973, which proved to be more effective than the natural compound. It was used by Vietnamese troops during the Second Indochina War and resulted in a significant reduction in mortality due to malaria.

Artemisinins are antimalarial agents with a good safety profile and combined artemisinin-based therapies are recommended by WHO as the first-line treatment for falciparum malaria in all malaria endemic areas. It replaces failing treatments (such as chloroquine) and reduces morbidity and mortality associated with malaria. The emergence of resistance to artemisinin appeared in 2008 at the Thai-Cambodian border, a historical zone of resistance to antimalarial drugs [DON 09].

Ariey *et al.* [ARI 14] identified mutations, called K13, in *P. falciparum* which was associated with delayed parasite clearance after treatment with artemisinin in Asia. The extent of resistance to artemisinin appears to have decreased in some areas of Southeast Asia, but is increasing in other parts of Asia and Africa [CUI 15].

6.6. Resistance to antibiotics

The evolution of resistance to antibiotics has become a global health emergency, as relayed by international organizations [WHO 14], with a tendency to increase their effects on human health [VAN 14]. However, for each new type of antibiotic, resistance can appear very rapidly [CLA 07] (Figure 6.5). Increased resistance to antibiotics in pathogens is a major public health threat with over 700,000 deaths per year [ONE 14].

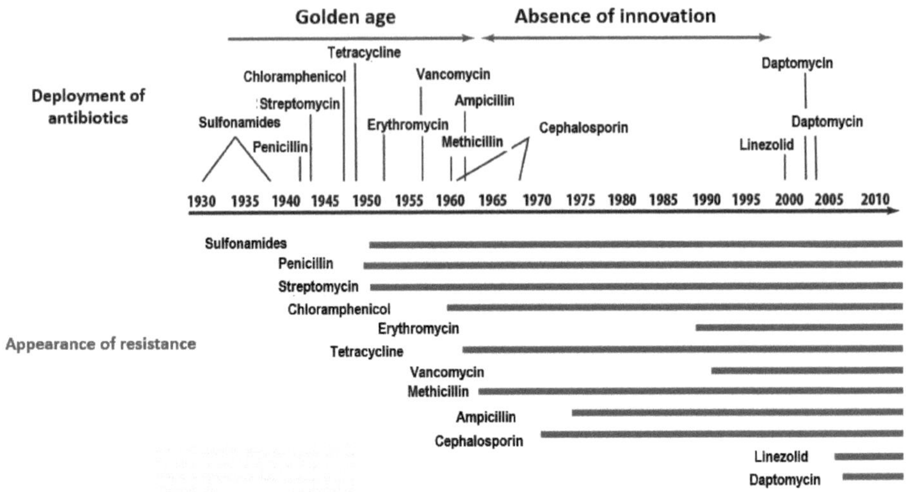

Figure 6.5. *Development and deployment of antibiotics followed by appearance of resistance (from [CLA 07]). For a color version of the figure, see www.iste.co.uk/morand/biodiversity.zip*

Resistance in environmental bacteria is nothing new. Antibiotic resistance genes or similar genes have been found in isolated environments with low anthropogenic impact, such as in bacteria isolated from permafrost dating back 30,000 years in Alaska [DCO 11] and in isolated caves [BHU 12]. Antibiotic resistance genes found in pathogenic bacteria originate from the environment [FOR 12, PAL 16]. Environmental bacteria are potential sources of clinically important forms of resistance and animals are potential reservoirs for these bacteria. Environmental resistance is still largely unexplored and little is known about the ability of environmental bacteria to conserve and transfer antibiotic resistance genes.

There are several antibiotic resistance mechanisms. Let us take the example of gram-negative bacteria [ALL 10]. Some of these are intrinsically resistant because they have an impermeable membrane or do not have the target for a given antibiotic (Figure 6.6). These bacteria have efflux pumps, which reject toxic compounds such as heavy metals or antibiotics to outside the cell. Some carriers such as the RND family (Resistance Nodulation Division) or the MFS (Major Facilitator Superfamily) can pump antibiotics directly outside the cell. Resistance mutations can modify the target protein by deactivating the antibiotic binding site while leaving the cell functionality of the protein intact. Inactivation of the antibiotic can be accomplished by modifying or degrading it.

In addition to these mechanisms for resistance through genes encoded on the bacterial chromosome, other mechanisms ensure that they are exchanged between bacterial organisms. These transmission mechanisms provide the host cell and its progeny with new genetic material, encoding resistance [HOL 16] (Figure 6.6). There are three main routes that ensure this transfer of genetic material: transformation, where bacteria are able to absorb free DNA from the environment and incorporate it into their chromosome; transduction, where the resistance genetic material is transferred between bacteria by bacteriophages (bacteria virus); conjugation, which is a mechanism of gene transfer through plasmids, with the formation of a tube between two bacterial cells allowing plasmid transfer. Thus, it was a colistin resistance gene carried by a plasmid that emerged in Asia and spread globally (Box 6.2).

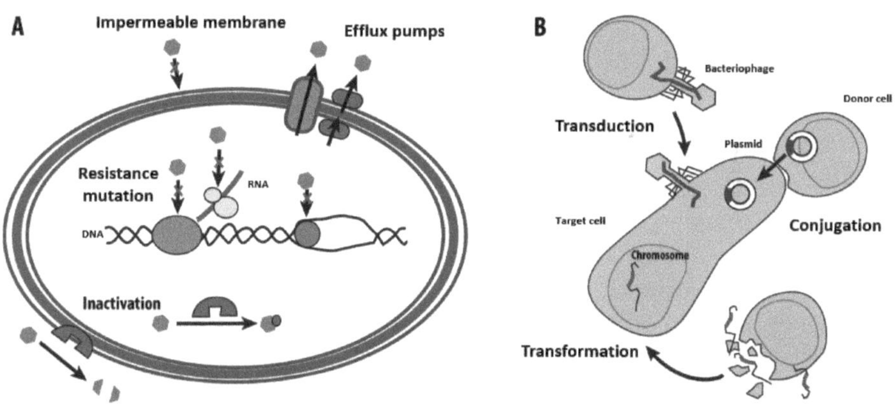

Figure 6.6. *Mechanisms of resistance: (A) and transfer of resistance genes; (B) in gram-negative bacteria. (A) Antibiotic resistance mechanisms are caused by: (i) membranes that are impermeable to certain antibiotics (or not targeted by the antibiotic); (ii) efflux pumps, which reject antibiotics (RND, Resistance Nodulation Division or MFS, Major Facilitator Superfamily); (iii) resistance mutations that disable the antibiotic binding site on the target protein; iv) inactivation of the antibiotic by modification or degradation (from [ALL 10]). (B) Genetic material is transferred between bacteria through three main pathways: (i) transformation, with environmental DNA recombination; (ii) transduction, where the genetic material is transferred by bacteriophages; (iii) conjugation with plasmid transfers (from [HOL 16]). For a color version of the figure, see www.iste.co.uk/morand/biodiversity.zip*

Subsequent to monitoring antibiotic resistance in the commensal bacterium Escherichia coli from living animals in China, [LIU 15] proved the emergence of a single plasmid carrying a colistin resistant mutation (mcr-1) in animals from all over China, as well as in human patients. This plasmid carrying mrc-1 was found throughout Southeast Asia, then in Europe and finally across the whole world [OLA 16, FER 16, MAR 16a]. In China and Southeast Asia intensive animal production systems coexist with smaller mixed systems including integrated aquaculture-livestock systems. These production units are generally characterized by inefficient waste treatment and widespread use of antibiotics in aquaculture resulting in high concentrations of antibiotic drug residues in wastewater. Antibiotics are also present in the feed for farm animals because of their effect as a growth factor. This latter use could explain the selection and subsequent passage of colistin resistance of pig commensal bacteria to human commensal bacteria, as observed in a village in northern Laos [OLA 15]. The fact that the plasmid carrying this mutation, as well as other types of resistance mechanisms, was found in villages with low agronomic intensification calls into question the respective shares of local conditions and regional circulations in the emergence and diffusion of antibiotic resistance.

Box 6.2. *Emergence of a plasmid carrying the mrc-1 mechanism of resistance to colistin*

Experiments have shown that antibiotic resistance genes have been increasing in abundance and diversity in the microbiome of pigs that are fed with antibiotics. Some of these enriched genes confer resistance to non-administered antibiotics, demonstrating the potential for indirect selection of resistance to non-administered antibiotic classes [LOO 12].

The study of mechanisms of antimicrobial resistance emergence is complicated by the high number of transmission routes for antibiotic resistance genes between environmental and commensal bacteria in animals and humans. Thus, antimicrobial resistance genes derived from livestock manure bacteria can affect environmental bacteria [UDI 14], which are also an immense reservoir of resistance genes [MAR 09a, THA 16] (Figure 6.7).

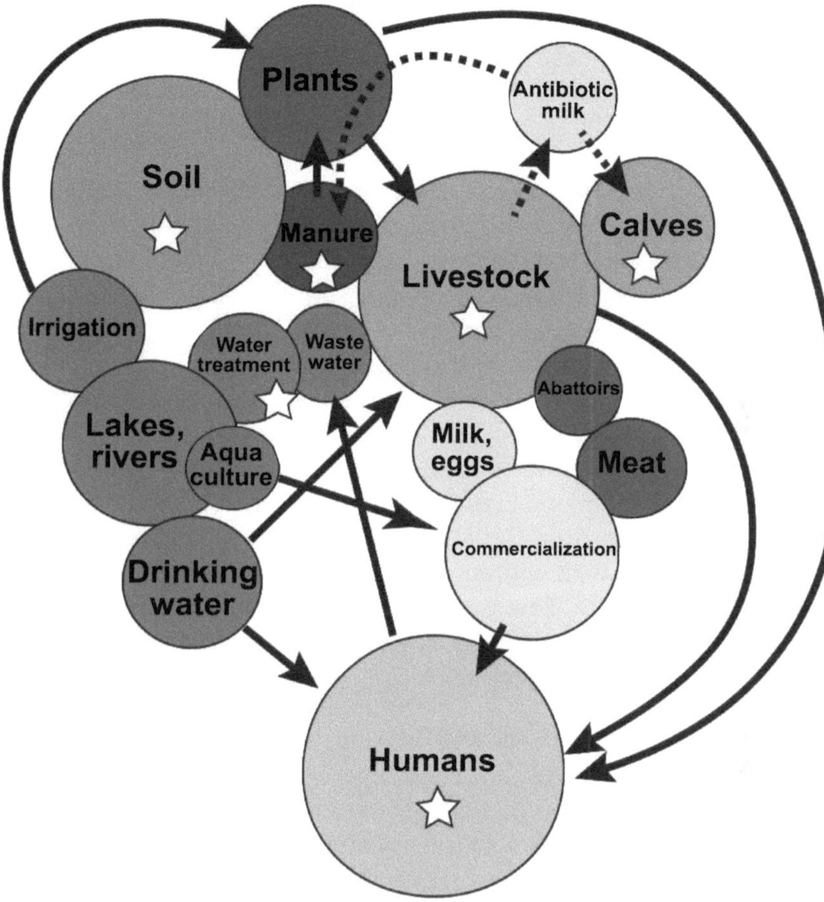

Figure 6.7. *Diffusion of antibiotics and antibiotic resistance genes in agriculture, the environment and the agri-food industry. The different colors define different groups of reservoirs. Stars indicate the hot spots of antibiotic resistance genes that depend on high bacterial densities and selective pressure, as induced by antibiotics in livestock and human metagenomes, manure storage facilities, sewage treatment plants and the rhizosphere (from [THA 16]). For a color version of the figure, see www.iste.co.uk/morand/biodiversity.zip*

Some insects are linked to humans and animals through their synanthropy and thus provide a link for transmission of bacteria that carry antibiotic resistance genes between livestock and urban communities. Flies and cockroaches bear bacterial clonal lines identical to the bacteria found in animal manure [ZUR 14]. Synanthropic rodents also play a role in the circulation of antimicrobial resistance. Thus, the incidence of antimicrobial resistance in Vietnamese farms in the Mekong Delta was linked to the presence of antimicrobial resistance genes in commensal bacteria from commensal rodents living near these farms [NHU 15].

A recent study [PAL 16] characterized the microbial taxonomic diversity, as well as resistance genes and their mobile genetic elements, metagenomes (bacterial population genomes) from humans, animals and various environmental sources like waste water, soil or smog. The study showed a low taxonomic diversity of human and animal microbial communities and their genes for resistance to biocides and heavy metals. On the other hand, metagenomes showed a high abundance of antibiotic resistance genes. Environmental metagenomes had a high taxonomic diversity with a high diversity of biocide and metal resistance genes. An important aspect is the high diversity of antibiotic resistance genes, including last resort antibiotics, in the metagenomes from Beijing's smog particles. This study showed that air can be polluted with antibiotics and is therefore a transmission route for antibiotic resistance. The study also highlighted the importance of environmental metagenomes as reservoirs of resistance genes for many unknowns that can be transferred to pathogens.

Heavy metal pollution as well as other environmental stressors such as biocides and pesticides can all contribute to the emergence and increase of resistance to antibiotics through cross-resistance mechanisms [SEI 12]. Ultimately, the systematic consequences of antibiotic use may extend to the production of greenhouse gases [HAM 16a] (Box 6.3).

Hammer *et al.* [HAM 06] studied the collateral effects of antibiotics on the microbiota of non-target animals associated with manure, such as beetles. Cattle were treated with a broad-spectrum antibiotic that is commonly used in animal husbandry and the effects were measured in manure microbiota and beetles. Antibiotic treatment thus deeply restructured the beetle microbiote, which housed a microbial community that was distinct from those of the manure they consumed. The authors also quantified greenhouse gas fluxes from manure. Antibiotic treatment has increased manure methane fluxes, probably due to a change in the interactions between archaea and methane bacteria in the rumen of cattle and manure. By modifying greenhouse gas emissions from manure, antibiotic treatment can have unanticipated ecological effects that far exceed the target animal.

Box 6.3. *Antibiotics and greenhouse gases*

6.7. Evolution of virulence

The virulence of a pathogen must be considered in the context of a compromise hypothesis, which assumes that there are both benefits and costs associated with virulence [FRA 96, VAN 98]. The cost is loss of transmission due to death of the host whereas the benefits of high virulence are increased transmission, for example increased production of infesting stages or viremia. Virulence results from a dynamic optimization between the cost and associated benefit. The virulence of pathogens is a result of adaptation to co-evolving interactions with their hosts in specific ecological environments. The virulence of a pathogen is therefore sometimes difficult to predict due to the complexity of processes that affect the pathogen's infection of the host and the environment in which the host and the pathogen live [KOE 01].

Acceleration of the genetic selection process of animals in industrial livestock sub-sectors is accompanied by a decrease in genetic diversity. Out of over 7,600 domestic breeds surveyed worldwide, over 20% are classified as endangered [FAO 15]. Pathogens in livestock and poultry live in environments with high host densities and very high genetic homogeneity. This high density of genetically uniform hosts favors viruses with high virulence [REA 01], which may explain the selection and amplification of highly virulent strains of avian influenza in industrial farms [CAR 08].

Natural immunity and vaccination do protect against pathogens, but pathogens can evolve by shortening their period of latency and potentially increasing their virulence. Thus, host populations with high levels of immunity may maintain pathogens that are more virulent than host populations with low levels of immunity [GAN 01]. Theoretically, immunization can select for virulence, as has been demonstrated by an artificial mouse screening experiment for high immunity against murine Plasmodium [MAC 04]. If host immunity can increase the selection for virulence of pathogens, then vaccines that reduce replication of pathogens may select for highly virulent pathogens [MAC 08a].

6.8. New biotechnologies and evolution of resistance: Wolbachia, CRISPR-Cas 9

In response to pesticide resistance, other approaches for vector control have been proposed by introducing sterile males. The sterile insect technique has been used successfully in limited and isolated areas such as Zanzibar (tsetse fly), Sanibel and Curacao (screwworm) or Okinawa [WEI 13a]. A variant of the sterile insect technique is the use of males infected with *Wolbachia* sterilizing bacteria. These intracellular bacteria have different cytotypes. They have the ability to cause breeding between infected male insects and uninfected females (or females infected by bacteria from another cytotype) to be sterile. *Wolbachia* bacteria increased the cost of insecticide resistance [DUR 06]. They also have the advantage of reducing the ability of insects to transmit pathogens such as viruses and as such, the use of *Wolbachia* was a promising approach in vector control [HOF 11].

Potential resistance to the effects of *Wolbachia* is still little known. However, a study showed that the West Nile virus is more in the *Culex tarsalis* mosquito in the presence of the *Wolbachia* wAlbB strain compared to mosquitoes that were not infected by this bacterium [DOD 14]. The amplification effect of the virus occurs because the *Wolbachia* wAlbB strain inhibits the antiviral immune route of Toll. The release into the environment of mosquitoes artificially infected with *Wolbachia* for vector control purposes requires a better understanding of the adaptive mechanisms and their evolutionary consequences.

Recently, CRISPR-Cas9 technologies have been proposed in many fields of application in human and animal health, pest and crop pest control, including control of biological invasions [WEB 15, HAM 16a]. The "Clustered Regularly Interspaced Short Palindromic Repeats" (CRISPR) sequences are DNA sequences from prokaryotes, like bacteria, which confer a form of acquired immunity to viral pathogens [SOR 08]. Research on this defense mechanism has provided a better understanding of the co-evolution between host bacteria and their bacteriophage viruses [VAL 10]. Bacteria can incorporate phage genetic material into their CRISPR sequences and this acquisition makes them resistant to phages that bear a sequence that is identical to the incorporated sequence. However, simple mutations in the phage genome are sufficient for the evolution of counter-resistance. An integral part of this anti-bacteriophage defense mechanism is the action of the Cas proteins associated with CRISPR. These Cas proteins are generally found adjacent to CRISPR sequences and show strong similarity to endonuclease proteins.

Thus, CRISPR-Cas9 is a system of endonucleases that intersect a target DNA sequence to complement a guide RNA (gRNA). The sequence can be inserted into a reading frame, and any mutations generated by CRISPR-Cas9 will produce copies of itself wherever the genome sequence is complementary to the gRNA, converting heterozygous individuals for the mutation into homozygous individuals. CRISPR-Cas9 offers such simplicity and precision in gene targeting that any standard molecular biology laboratory can develop this technology [WEB 15]. However, unwanted effects of the use of CRISPR-Cas9 technology have been observed in the control of the HIV-1 virus, which may lead to the emergence of Cas9-resistant viruses [WAN 16]. The high mutability of the viruses would be partly responsible for this resistance, effectively limiting the therapeutic use of the Cas9 technology.

6.9. Ecological and evolutionary engineering

The sciences of ecology and evolution provide recommendations for the use of biocides, pesticides and antibiotics. Research on the evolutionary dynamics of biological diversity (of genes and populations) finds its application in the development of eco-evolving engineering (Table 6.1).

Methods for slowing the evolution of resistance	Examples
Reduce the variability of a trait linked to the selective value	
Multiple drugs	Triple combination antiviral therapy of AIDS, antibiotic cocktail, "pyramiding" of pesticides
Ensure the effective dose	Tuberculosis therapy, antibiotic resistance CRISPR-Cas9
Reduce the appearance of resistance mutations	CRISPR-Cas9
Reduce the size of the target organism population	Integrated Pest Management
	Non-drug therapy
Reduce directional selection	

Vary selection over time	Herbicide rotation
	Variation of antibiotics, pesticides or antiretrovirals
Use non-chemical means of control	Integrated Pest Management
Limit exposure of target organisms to selection	Restrict the use of powerful (or last resort) drugs
Avoid broad-spectrum antibiotics	Sensitivity test to drugs or pesticides prior to use
Reduce the heritability of a trait linked to selective value	
Dilute resistance genes	Establishment of refuges

Table 6.1. *Evolutionary engineering and mechanisms that reduce the emergence and evolution of resistance*

6.9.1. *Management of resistance to Bt transgenic plants*

The emergence of resistance to genetically modified plants has made it possible to compare resistance monitoring data with the main predictions of evolutionary theory, such as refuge strategies and "pyramiding" of pesticides for the management of pest resistance in Bt crops. Resistance is less likely to change rapidly if Bt protein production is sufficient, frequency of the initial resistance allele is low, refuges are abundant and Bt plants possess two-toxin pyramids [TAB 13]. Abundant refuges delay the emergence of resistance, which has already been observed in the case of resistance to synthetic insecticides. Resistance evolves more slowly if plants are transformed to produce two toxins [ZHA 05]. Transgenes provide better control of the entire complex of target species as lesser toxin-sensitive species may be more sensitive to a second toxin. Ultimately, plants with Bt toxin pyramids require a smaller refuge. Theoretical models suggest that a 30–40% refuge is required when transgenes are used separately, whereas only 5–10% would be needed for pyrogenic plants expressing two transgenes [BAT 05].

6.9.2. *Managing antimicrobial resistance*

Alternating at least two classes of antibiotics over a period of several months to years is presented as a treatment strategy to slow down the evolution and spread of antibiotic resistance. An evolutionary model shows that this strategy is ineffective and suggests that antibiotic mixing strategies are much more effective [BER 04]. The ecological explanation for these results is that the mixed use of heterogeneous antibiotics slows down the spread of resistance.

As natural predators of bacteria, phages were used to treat bacterial infections prior to the advent of antibiotics, after which the use of therapeutic phages was restricted to the USSR [155]. With the increase in antibiotic resistance, phages have regained therapeutic interest due to their specificity of host and target bacteria, reducing the non-targeted effects on the rest of the microbiome. Active phages have been identified against several pathogenic bacteria, but as in the case of antibiotics, evolution of resistance to the phage has been seen [LAB 10, LU 11]. Adaptive values should therefore be better understood – of competition between bacteriophage-resistant mutant bacteria and phage-bacterial co-evolution [LEV 04, LAN 16].

6.9.3. *CRISPR-Cas9 technologies*

As pointed out by Webber *et al.* [WEB 15], promoters of CRISPR-Cas9 technologies must take into account the lessons learned from decades of research on the control of biological techniques. In particular, there is a need to better understand the mechanisms and specificity of CRISPR-Cas9 targets, their population-level consequences and the adaptive mechanisms that may lead to unwanted changes [UNC 17]. Ultimately, the unintended consequences must be examined on the scale of communities and environments such as ecological cascades.

6.10. Conclusion

The living organism that is to be managed or eradicated, such as crop pests, vectors and reservoirs of pathogens (and pathogens themselves) respond to the selection pressures of products used (pesticides, insecticides, biocides, vaccines, antibiotics) through adaptive mechanisms. Hypotheses derived from evolutionary theory show the importance of preventing or slowing down the evolution of resistances and unintended effects. Theory, modeling and empirical experiments are crucial for classical management methods and new technologies for the fine manipulation of life.

This chapter raises questions about what research best takes into account biodiversity and health relationships between reductive approaches that depend on high technicality, and integrative and systemic approaches. A second type of question concerns the qualification of antibiotics and other biocides such as pesticides, as "common goods" of humanity.

7

Animal and Human Pharmacopoeias

7.1. Introduction

Biodiversity is often presented as an essential source of pharmaceutical drugs and molecules in the agrochemical industry [WRI 00]. The total number of natural plant products was estimated to be more than 500,000 by Mendelsohn and Balick [MEN 95] with only 5,000 species of plants studied for their medical properties out of an estimated 350,000 to 400,000 species [ALV 14]. The systematic search for natural products in bacterial microorganisms and filamentous fungi was subsequent to the development of penicillin and the golden age of antibiotic discovery (Chapter 5).

Human societies have used plants and herbal products as medicines for millennia, and archaeological evidence supports the theory that medicinal plants were regularly used [HAL 05]. All ancient cultures and civilizations used plants or herbal preparations for curative or psychotropic purposes [PHI 01]. The first indications of the existence of phytotherapy came from a 60,000-year-old Neanderthal tomb that contained pollen and various medicinal plants [SOL 71]. More recently, among the effects of Ötzi (the Tyrolean Iceman), a modern man who lived 5,200 years ago and was discovered in 1991 in a glacier in the Italian Alps, were the remains of mushrooms and berries [CAP 98] (Box 7.1).

In 1991, a completely intact human body was discovered in a glacier in the Tyrolean Alps of Otztal, on the border between Italy and Austria. An analysis of the remains of this man, named Ötzi, revealed an age of 46 years at the time of his death, about 5,200 years ago. The presence of an arrow buried in his back testifies to a violent death. Paleo-ethnobotany or ethnobotany of past human societies looks at the relationships of these past men and their environments through the study of artifacts and visible remains of pollens, mosses, plants, animals and mushrooms. Examining the digestive tracts has revealed clues to the diet of men from that period [BOR 00]. Ötzi ate chamois, deer meat and cereals before his death. Over 80 species of bryophytes have been identified in his digestive tract. Ötzi carried many artifacts: an ax and bow made of yew tree (*Taxus baccata*), arrows made of viburnum wood (*Viburnum lantana*) and a dagger handle made of ash wood (*Fraxinus excelsior*). He had been carrying prunes (*Prunus spinosa*), birch bark (*Betula sp.*), pieces of charcoal in maple leaves (*Acer sp.*), and two species of fungi, *Fomes fomentarius* for tinder and *Piptoporus betulinus* for medicine.

Box 7.1. *Plants and Ötzi's pharmacopoeia (from [NOL 11])*

Most medicinal plants that are currently of interest come from the ancient civilizations of Africa, Asia and the Americas; in other words, the centers of the Neolithic Revolution, domestication of plants and animals and emergence of the first agrarian civilizations (Chapter 2, Figure 2.3). The use of medicinal plants is still important today and is recommended by the World Health Organization (WHO) [WHO 11]. The global market for herbal medicines was worth at least $19.5 billion in 2008 with an annual growth rate of 11.0% (International Trade Center, http://www.intracen.org/itc/sectors/medicinal-plants/). This trade involved over 3,000 species. India has about 2 million traditional medical practitioners using over 7,500 species of medicinal plants [SAL 15]. Ethnobotany has been established as a discipline that seeks plant sources and their bioactive molecules in traditional pharmacopoeias. Old pharmacopoeias are also being rediscovered, as in the case of artemisinin through the "Handbook of Prescriptions for Emergencies" by the physician Ge Hong (see Chapter 6).

In this chapter, we explore the origin of self-medication in animals and the links between animal and human pharmacopoeias. We will then address the diversity of ethnobotanical knowledge and traditional medicines before discussing the threats to this knowledge and the associated biological diversity.

7.2. The diversity of plant secondary metabolites

Plants produce a wide variety of molecules of low molecular weight. As already mentioned, over 50,000 of these metabolites have been structurally studied [MEN 95], but the actual number of existing metabolites range in hundreds to thousands. Some of these metabolites involve metabolic pathways that are common to all organisms but the vast majority of others are called "secondary" metabolites, a historical term that was initially associated with a lack of recognized utility for these plant metabolism products. As noted by Theis and Lerdau [THE 03], our modern appreciation of the importance of these secondary metabolites from an ecological and evolutionary perspective originates from Fraenkel's article [FRA 59] "Rationale for the secondary plant substances", whereby he suggests that the majority of chemicals found in plants are not without utility but are metabolism byproducts with important functions (Figure 7.1). A "secondary" metabolite is defined as a compound for which the biosynthesis is limited to a restricted group of plants [PIC 00].

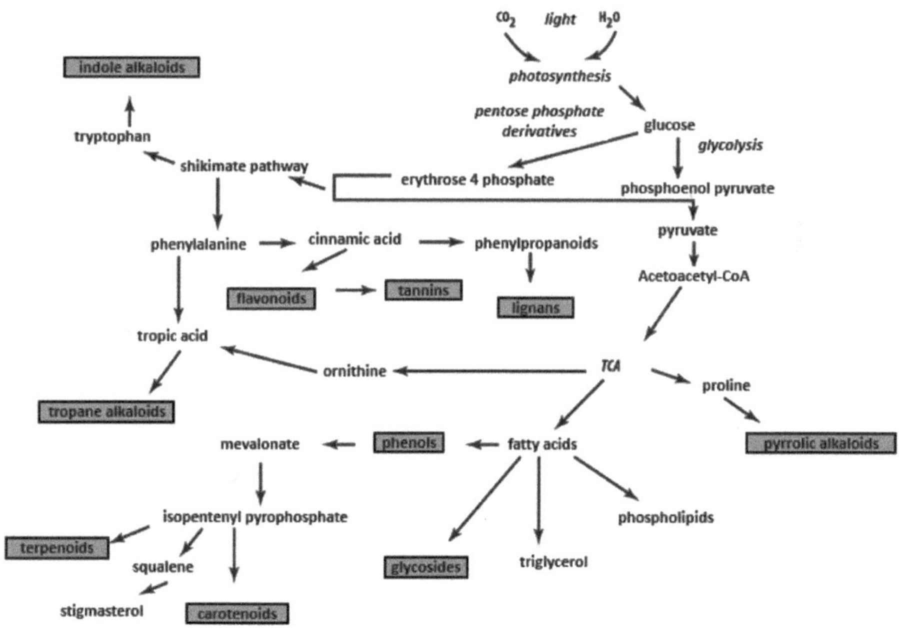

Figure 7.1. *Primary and secondary metabolism of plants, the main types of metabolites are in boxes (from [ALV 14]). For a color version of the figure, see www.iste.co.uk/morand/biodiversity.zip*

The ability to synthesize secondary metabolic compounds has evolved over the diversification of plant lines. Volatile substances and pigments have evolved in response to pollinators (to increase fertilization rates), toxic substances (in response to pathogens like bacteria and fungi), predation of herbivores (insects, vertebrates) and competition with other plants. Metabolites in fruits may slow down their deterioration or act as signals to attract animals that will then disperse seeds after ingesting the fruits [PIC 00]. These metabolites involve chemical interactions between plants and animals and are the subject of a specific discipline: chemical ecology [EIS 95, DIC 06].

Many metabolites have medical properties. Pharmacognosy is the study of drugs derived from plants or other natural sources. The American Society of Pharmacognosy defined this discipline as "the study of the physical, chemical, biochemical and biological properties of drugs, drug substances, or potential drugs or drug substances of natural origin as well as the search for new drugs from natural sources"[1].

These metabolites include alkaloids that act on a wide variety of metabolic pathways in humans and other animals (they almost uniformly have a bitter taste); glycosides, which are molecules that contain a sugar bound to a non-carbohydrate moiety and terpenes, which are a broad and diverse class of organic compounds (such as turpentine).

Methods for identifying and characterizing the biological activity of metabolites include fractionating the crude extract from a plant by chromatography and using solvents to test the fractions *in vitro* until pure active compounds are obtained. The structures are then determined using spectroscopy [ALV 14]. Many metabolites have been characterized depending on their actions on human health (Table 7.1).

Developments in metabolic and genetic engineering and synthetic biology have increased the quantity and quality of relevant secondary metabolite biosynthesis.

1 http://www.pharmacognosy.us/what-is-pharmacognosy

Metabolites	Plant sources	Biological activities
Terpenes		
Caryophyllene	*Cannabis sativa*	Anti-inflammatory
Santonin	*Artemisia maritima*	Antiparasitic
Artemisinin	*Artemisia annua*	Antimalarial
Taxol	*Taxus brevifolia*	Anti-cancer
Digitalin	*Digitalis lanata*	Cardiotonic
Alkaloids		
Atropine	*Atropa belladonna*	Anticholinergic
Cocaine	*Erythroxylum coca*	Anesthetic
Quinine	*Cinchona ledgeriana*	Antimalarial
Vinblastine	*Catharanthus roseus*	Antineoplastic Anti-cancer

Table 7.1. *Biological activities of different terpenes and alkaloids derived from plants (from [ALV 14])*

7.3. Origin of self-medication in animals and hominids

The use of plants for therapeutic purposes can be found in the evolution of interactions between animals and plants. Many animal species show the capacity to self-medicate through plants [LOZ 98], including domestic animals [VIL 06]. Non-human primates are able to recognize and use different plants for their healing properties [HUF 01a, KRI 05]. Observing the behavior of animals in choosing and consuming plants is a source of understanding the evolution of behavioral defenses in humans as well as a source of knowing new natural products for maintaining health.

Janzen [JAN 78] was the first to hypothesize that herbivores benefit from access to secondary metabolites in plant food for their health. Some animal species actively intake specific plants with metabolites having analgesic, antimicrobial, anti-inflammatory or antidiarrheal properties [HUT 03, WRA 95]. Birds and ants collect foliage or resin for their antiparasitic or antimicrobial properties [CLA 88, CHR 03a]. Primates also consume soil and clay (geophagy) for sanitary purposes [KRI 00]. These behaviors are difficult to detect and interpret [ENG 02]. One of the difficulties in interpreting self-medication lies in the distinction between health benefits and nutritional benefits that are derived from the consumption of plants rich in secondary compounds [DER 13]. Thus, if correlation studies are compatible with the self-medication hypothesis, they do not establish a causal link [LOZ 98].

An interesting aspect of self-medication from an evolutionary point of view is the immune response. Since investment in immune responses is generally quite costly, drugs offer an alternative to cellular and humoral immune responses and limit the activation of different immune pathways [PAR 11]. African primates and apes are the most studied animals, in particular, for associating self-medication behavior with control of intestinal parasitic rates of infection [HUF 96a, HUF 01a]. Behavioral ecologists and especially primatologists have shown interest in animal self-medication and its potential applications for modern medicine. Animal self-medication or zoopharmacognosy is the study of animal behaviors in the selection and ingestion of plants, but also of clay, soil or insects to reduce the harmful effects of parasitic infections or toxins [HUF 97]. One of the best examples of self-medication is consumption of the plant *Vernonia amygdalina* by wild chimpanzees suffering from parasitic diseases, although this plant is not part of the usual diet of these animals.

Domestic animals also exhibit self-medication behaviors [HOS 10]. Lambs can detect the presence of internal parasites, or symptoms associated with parasitism and modify their diets by ingesting plants that are rich in antiparasitic agents, such as tannins [VIL 07, VIL 09, LIS 09].

In studying animal self-medication, the question is raised on mode of behavior acquisition, the choice of medicinal plants and the way in which individuals become predisposed to ingesting medicinal plants. One of the questions posed by primatologists relates to the existence of medical traditions and the role of innate and learned self-medication. Hart [HAR 05, HAR 11] suggested that animals have an innate tendency to choose appropriate plants when they feel sick. The use of medicinal plants and their specific parts to prevent or treat certain diseases would be the result of a natural selection for a behavioral predisposition to seek and use plant elements. According to Hart [HAR 11], learning is an unlikely explanation for most aspects of acquiring the practice of herbal medicine. Hominids, the ancestors of present-day humans, would not have learned aspects of medicine by observing non-human animals but, subjected to the same selective forces in the fight against pathogens and parasites, evolution would have led them to strategies of defense and behavior similar to other animals [HAR 05].

Great apes, gorillas, chimpanzees and humans select the same plants for similar diseases or injuries because of their phylogenetic proximity as well as similarity in physiological perceptions of symptoms and their cognitive abilities to associate and select similar properties in medicinal plants [HUF 96a]. Thus, similarities can be seen between chimpanzees and humans in their use of *Vernonia amygdalina* and its effectiveness in controlling internal parasites [HUF 96b]. Our hominid ancestors show similarities with current primates and modern humans in the selection criteria of plants. This hypothesis is supported by Krief *et al.* [KRI 15] for whom chimpanzees are a model to understand the medicinal practices of Neanderthals.

However, humans are different because of two behaviors: the preparation of medical compositions and practices of care given to their fellow humans. Several African ethnic groups have developed a medicinal composition based on *V. amygdalina* for the treatment of various conditions such as malaria, schistosomiasis, amoebic dysentery, intestinal parasites and stomach pains [HUF 96a]. In addition to these cultural practices studied by ethnomedicine is the observation of animals. Primatologist Michael Huffman [HUF 01b] told the story of a Tanzanian doctor who discovered an important treatment nearly a century ago that saved the lives of many people suffering from a dysentery epidemic in his village. This doctor learned about the potential medical value of a plant by observing a young porcupine struck by the same disease that was consuming its roots. The inhabitants of the village avoided this plant, which they knew to be toxic, but after telling his story about the porcupine to the villagers, the Tanzanian doctor persuaded them to use the plant. Hoffman [HUF 01a] learned this story from the doctor's grandson, himself a respected doctor who continued to use this plant to treat different infections. From their evolutionary history, humans have inherited a behavioral predisposition to seek and use plant elements, says Hart [HAR 11]. On the other hand, they have developed a rich pharmacopoeia and medical knowledge that evolves and is transmitted culturally. Thus, pastoralists and pastoral societies have developed ethno-veterinary knowledge on livestock diseases and useful medicinal plants. An ethnobotanical study conducted in a pastoral society in Uganda listed 209 plants used for traditional veterinary health [GRA 09a].

7.4. Ethnobotany and traditional medicine

The term "ethnobotany" was created in 1895 by the botanist Harsherberg. He defined it as "the use of plants by primitive and aboriginal peoples". In 1941, Jones defined it as "the interrelationships of primitive men and plants" [ALV 14]. Ethnobotany plays an essential role in the discovery of plant sources and bioactive molecules. Its current definition has become "the study of how people of a particular culture and region make use of indigenous (native) plants"[2].

The origin of this science lies in the colonization of the non-European world and Europeans searching for new medicines to combat tropical fevers [BOU 16]. In the Americas, at the beginning of its colonization, the Spanish monarchy launched investigations to collect botanical knowledge from the Amerindians. Ethnobotany then developed with Jesuit missions and compilation of the first floras of American plants and their native medical uses. During the French colonial period, French botanists carried out research on the taxonomy of medicinal flora in Vietnam, which was published in the "General Flora of Indo-China" (Lecomte 1912–1937). Like the French, Spaniards nevertheless denigrated the native populations' medical knowledge. Thus, despite traditional medicine dating back to at least the 2nd Century BC and a pharmacopoeia that was rich in over 3,800 traditional remedies Le Tran Duc [LE 95], quoted in Ngoc Van [NGO 08]), the French colonial regime did not encourage the use of traditional medicine nor its inclusion in the colonial health system.

Over the past 40 years, the methods and goals of ethnobotany have profoundly changed and expanded [NOL 11]. In addition to initial questions about plant uses for medical (and for colonial) purposes, questions have evolved towards perceptions of diseases and the symbolism of treatments, how and why decisions are made about where and when to collect medicinal plants and the effects of anthropogenic ecological changes on the pharmacological potential of plants [ETK 11]. Ethnobotanical research requires immersion into the local culture and its links to the ecosystem. Many ethnobotanical studies use participatory observation techniques enabling ethnobotanists to adopt the daily lifestyles of the people they want to learn from [NOL 11]. This involves taking part in daily activities such as housework, water collection, help in the garden, fishing, hunting or gathering and preparing food with community members. Discursive and

2 USDA https://www.fs.fed.us/wildflowers/ethnobotany/.

narrative methods (note-taking and tapping) are used in the observation of participants for qualitative analysis [WIL 04]. Cunningham [CUN 01] recommended verifying the names of plants with different people and comparing the results by presenting specimens of plants to participants for identification and use, or at local markets where participants were asked to name and give the use of plants. Herbariums are also necessary for modern taxonomy and museum reference collections.

Local practices for managing resources and medical plants are based on knowledge of the local environment called "Traditional Ecological Knowledge" (TEK), which combines techniques and tools for managing local ecosystems [NOL 11]. These practices are guided by social institutions and shaped by visions, and therefore should not be taken out of context [BER 08]. Participatory ecological research provides an essential, yet underused tool for documenting the ecological consequences of local management practices [NOL 11]. Anthropogenic changes, such as changes in land use, as well as market entry and demand for plant and pharmaceutical resources, all affect the biological diversity of medicinal plants and indigenous and traditional knowledge.

7.5. Bioprospecting, biopiracy and patents

Biodiversity is the fundamental resource for bioprospecting; in other words, the collection, analysis and commercialization of genetic resources. It may also include the collection of traditional knowledge for the use of these resources. Sharing the benefits of commercialization with countries that have these genetic resources must be ensured through bioprospecting, and if indigenous knowledge is used in the development and commercialization of a commodity, sharing benefits with local communities is imperative. When traditional resources or knowledge are used for commercialization and the benefits are not shared with the concerned parties, indigenous communities and countries, it leads to biopiracy [SAL 15].

In 1988, the first International Congress of Ethnobiology called for fair compensation and legal defense of the knowledge of indigenous peoples: the Declaration of Belem (Box 7.2). The Manila Declaration, which was published in 1992 by a network of Asian botanists and pharmacognosists sponsored by UNESCO, provides ethical guidelines for the use of Asian biological resources. An appendix to this statcment includes specific guidancc for biological sample collectors (see [BOD 07a]).

As ethnobiologists, we are alarmed that:

SINCE

– tropical forests and other fragile ecosystems are disappearing;

– many species, both plant and animal, are threatened with extinction;

– indigenous cultures around the world are being disrupted and destroyed;

AND GIVEN THAT

– economic, agricultural and health conditions of people are dependent on these resources;

– native people have been stewards of 95% of the world's genetic resources, and

– there is an inextricable link between cultural and biological diversity;

WE, MEMBERS OF THE INTERNATIONAL SOCIETY OF ETHNOBIOLOGY STRONGLY URGE ACTIONS AS FOLLOWS:

1) Henceforth, a substantial proportion of development aid must be covered to efforts aimed at ethnobiological inventory, conservation and management programs.

2) Mechanisms should be established by which indigenous specialists are recognized as proper authorities and consulted in all programs affecting them, their resources and environments.

3) All other inalienable human rights be recognized and guaranteed, including cultural and linguistic identity.

4) Procedures must be developed to compensate native people for utilization of their knowledge and biological resources.

5) Educational programs must be implemented to alert the global community about the value of ethnobiological knowledge for human well-being.

6) All medical programs include the recognition of and respect for traditional healers and the incorporation of traditional health practices that enhance the health status of these populations.

7) Ethnobiologists make available the results of their research to the native people with whom they have worked, especially including dissemination in their native language.

8) Exchange of information should be promoted among indigenous and rural people regarding conservation, management and sustainable utilization of resources.

Box 7.2. *The Declaration of Belem at the First International Society of Ethnobiology Congress 1988*

Two international legal frameworks, the Convention on Biological Diversity (CBD) and intellectual property rights systems that touch on the WTO's Agreement on Trade-Related Aspects of Intellectual Property Rights (TRIPS) converged in the Doha Declaration in 2001. The Doha Declaration addresses intellectual property rights and patenting aspects related to the protection of traditional knowledge and folklore (Box 7.3).

The Nagoya Protocol, which was adopted in 2010 and entered into force on October 12, 2014, supplemented the provisions of the Convention on Biological Diversity, which states that each State is sovereign over the genetic resources in its territory and over any associated traditional knowledge.

Following the Bonn Convention on Biodiversity, the Bonn guidelines were adopted by an *ad hoc* working group (ABS CBD COP 5 Nairobi 2000) to clarify the steps to be followed in the process of access and sharing of benefits as stated by the CBD.

The Nagoya Protocol was adopted subsequent to the Johannesburg Summit in 2002, which found that the benefit-sharing provisions of the CBD were not being respected. It aimed at meeting the third objective of the CBD – fair and equitable sharing of benefits arising from the use of genetic resources and combating biopiracy. Let us note that the Protocol also applies to use of both genetic resources and associated traditional knowledge held by indigenous and local communities.

The Protocol contains obligations for both user countries and countries providing genetic resources: the former must ensure that they comply with the ABS legislation in force in the supplier countries (Article 15) while the latter must ensure that its legislation implements the international provisions of the Protocol in a clear and transparent manner (Article 6.3).

Thus, potential users of genetic resources must obtain Prior Informed Consent (PIC) from the country in which the resource is located before access is granted.

This consent may, depending on the country, involve obtaining a permit, for example.

Access to and use of the resource is also subject to negotiation and Mutual Agreed Terms (MAT), which should include and detail how benefits resulting from the use of the resource will be shared with the provider, as a prerequisite for access to the resource and its use. The formalization of this agreement depends on national legislation and can sometimes be incorporated into the PIC (on formalization, see [MOR 14a].

It should be noted that the rights of local and indigenous communities must be taken into account (Article 12 of the Protocol) in all cases and cannot be subordinated to their recognition by national law [MOR 14a, pp. 170–177]. This may result in community prior consent (Community PIC).

The CBD also proposed the development of *sui generis* protection systems, which include measures to preserve and ensure respect for and promotion of the knowledge, innovations and practices of indigenous and local communities. These practices embody traditional lifestyles that encompass sustainable use of biological diversity, including associated biological and genetic resources (UNEP/CBD/WG8J/5/6, 2007). A number of these systems have already been developed [DAG 14].

For matters relating to national regulations on the implementation of the CBD and the Nagoya Protocol, the national focal point[3] should be approached.

Box 7.3. *Access and benefit-sharing tools*

Legal and ethical discussions have taken place in recent years on the intellectual property of traditional knowledge. Dutfield [DUT 00] considered that patents are not a viable option for most indigenous communities or traditional knowledge repositories, as knowledge is collectively held and maintained, while patent law refers to inventiveness and the realization of individuals. Dutfield [DUT 00] also pointed out that patent applications must provide evidence of a single act of discovery.

3 https://www.cbd.int/information/nfp.shtml

The Declaration of Belem, along with other declarations and agreements, led to the signing of the Nagoya Protocol 2010, which complemented international legal rules related to access and sharing of biodiversity resources and knowledge (see Chapter 11).

7.6. Conservation biology and traditional pharmacopoeia

The surge in global demand for herbal medicines was accompanied by a decline in the supply of medicinal plants worldwide. Overexploitation for commercial purposes, destructive harvesting practices, habitat losses linked to forest degradation and agricultural hold-up were contributing factors to these threats of medicinal plant diversity [BOD 07a]. Such an ecological impact is not recent; the overexploitation of guaiac wood and cinchona wood has been observed since the beginning of the Spanish colonial era in South America, which led to the implementation of resource control measures by colonial authorities. The impact of overexploitation of these woods are still visible today. For example, all Guaiacum species including guaiac wood trees are on the Convention on International Trade in Endangered Species' (CITES) wild fauna and flora list and are subject to protection and international regulation of their trade.

Cunnigham [CUN 93] showed that for Africa, urban growth was accompanied by an increase in urban markets for traditional medicines placing pressure on rural stocks. The supply of plants for traditional medicine can no longer meet the demands of urban areas while access to biomedicine is still difficult for the vast majority.

The Chiang Mai Declaration in 1988 emphasized that medicinal plants are essential in healthcare, both in terms of self-medication and for national health services, and expressed concern about the consequences of loss of plant diversity. It mentioned "the urgent need for international cooperation and coordination to establish medicinal plant conservation programs in order to ensure that adequate supplies are available to future generations" [AKE 91].

In Vietnam, many traditional medicines use plant and animal resources that are classified as endangered. A survey carried out by the NGO TRAFFIC (http://www.traffic.org/, Ngoc Van *et al.* [NGO 08]) on local markets found that half of the 68 animal species sold on markets, whether still alive or as by-products, are listed on the IUCN Red List of Endangered Species. Some of these animals that are used in traditional Vietnamese medicine are hunted, bred in captivity, or imported legally or illegally from neighboring countries (mainly China, Cambodia, Laos and Malaysia). Other animals are hunted as youngsters and kept in captivity for later sale. This list includes carnivores and many turtle species. The majority of plants sold comes from natural harvesting and, among them, 27 species are considered to be endangered. The survey revealed that sellers were generally unaware of conservation problems of the species they sold. The impact of traditional medicines on conservation of flora and fauna poses both legal and ethical problems. Thus, a traditional medicine can only be banned on the basis of its danger to health. A pharmacopoeia that does not have medical properties proven by biomedicine, such as rhinoceros horn powder, cannot be banned from preparation and sale.

7.7. Loss of biodiversity and knowledge

Loss of biodiversity is accompanied by loss of cultural diversity (Chapter 3, Figure 3.6). A study conducted in Yunnan in China showed that people were no longer able to name locally extinct bird species [KAI 14]. If people were still able to identify the species of locally extinct mammals, it was because they knew them through television documentaries. Young people no longer know forest animals except through cultural memory and not through vision or sound anymore [KAI 14].

This loss of knowledge on biological diversity also concerns medicinal plants [RAM 07]. Knowledge of medicinal plants is particularly vulnerable due to our growing dependence on biomedical care, urbanization and the loss of interest in the younger generations in botany and traditional knowledge [CAS 05, BEN 00]. While some studies show that this knowledge can be maintained in first-generation immigrant populations [VAN 12], it seems that biodiversity loss, habitat loss and cultural changes lead to a loss of botanical knowledge with direct consequences on population health [MCD 07].

7.8. Conclusion

This chapter explored the influence of diversity in secondary plant metabolites on the evolution of behavioral defenses in animals and humans.

The developments of traditional and modern pharmacopoeias have enriched medications available to treat many health problems.

We have seen that the quest for knowledge from indigenous people by European colonizers has contributed to the enrichment of botanical knowledge through ethnobotany, but often to the detriment of indigenous and traditional knowledge. Bioprospecting and the search for new molecules are still accompanied by unethical practices, even biopiracy, despite the improvement in international legal frameworks.

Radical critics consider that science has served the interests of colonial development and European imperialism of the 19th Century, but that new forms of appropriation of resources and knowledge are now at work [WHI 09]. The term "biocolonialism" has been put forward to determine this new relationship of appropriation of biological resources and knowledge by means other than under direct government control. The first step would be to declare that indigenous genetic resources belong to all, as "common heritage of mankind", to then be exchanged as a "free good". These genetic resources would not belong to any people because they are the product of evolution and past human heritage. These free "public" goods could then be converted into private property using the intellectual property rights system [WHI 09].

We have seen that the current loss of biodiversity is accompanied by a loss of cultural diversity and local knowledge of animals and plants, including medicinal plants. Simultaneously, the demand for certain natural products has created new threats, both to species and landscapes but also on traditional knowledge [MAF 10].

Functionally, loss of biodiversity results in the loss of an extensive network of secondary metabolites that are essential for interactions with animals and humans, such as metabolites with antiparasitic and antimicrobial properties. The simplification of agricultural systems to meet agro-industrial production that widely uses pesticides, biocides and antibiotics (Chapter 6) is accompanied by a reduction in the diversity of secondary metabolites. As Villalba and Provenza [VIL 07] pointed out, integrating animal self-medication into the design of sustainable grazing systems would both improve and protect biodiversity, soil quality, and health and well-being of animals and ultimately of humans.

8

Well-being

8.1. Introduction

Well-being is not just the absence of disease. The World Health Organization (WHO) defines human health as "a state of complete physical, mental and social well-being" (WHO 1946) (see Chapter 1). The Millennium Ecosystem Assessment (MA) [MA 05] developed a conceptual framework that connects ecosystem services provided by biodiversity with human well-being (Chapter 1). This ecosystem assessment defines human well-being through security, access to basic resources, health, good social relations and freedom of choice for individuals. This definition includes the objective and subjective characteristics of well-being. Contrary to what we have seen in the previous chapters in this book, the MA [MA 05] does not distinguish between the biological, evolutionary, psychological and epidemiological aspects of human health and well-being. The same can be said for the definition of well-being and human health given by IPBES.

This chapter explores the links between well-being and biodiversity, looking for its evolutionary origins and its most current socio-cultural features. The notion of biophilia is central to this topic [WIL 84], but it is not the only conceptual framework, especially for psychologists. To begin with, let us consider the objective and subjective definitions of human well-being.

8.2. Objectivity and subjectivity of well-being

According to Diener [DIE 09], well-being is defined either objectively, for example, the Human Development Index (HDI), or subjectively, so according to the interests of the person whose well-being is being assessed. People often express a good level of well-being when they feel that their lives are going well. This subjective conception of well-being comes close to the concept of utility of economists as the result of maximization of personal happiness. Diener [DIE 09] also stressed the need to take into account the affective responses of people, in order to know whether the conditions of their environment correspond to their preferences or not.

McGillivray [MCG 07] summarized many conceptualizations of well-being and stressed that the term is an abstraction that refers to everything that can be assessed from a person's life situation. This situation corresponds to the quality of life, the standard of living and the level of personal development of the person. Other notions of well-being can combine social protection, life satisfaction, prosperity, autonomy and happiness. The fulfillment of basic human needs is also a component of well-being.

Seymour [SEY 16] added the concept of social health defined as "the ability to lead life with some degree of independence and participate in social activities" to social well-being (see [HUB 11]). Indicators of well-being must take social cohesion and participation in community activities into account, in order to show that physical and mental well-being are dependent on the social capital of individuals [ZIE 05]. Finally, another conceptualization of well-being is that of "capabilities" by Amartya Sen [SEN 85]. Thus, well-being is best considered in its multiple dimensions, which cannot be reduced to a single dimension and, as Alkire ([ALK 02], cited in [MCG 07]) pointed out, without implementing a hierarchy among the various dimensions of well-being.

Objectively, well-being can be valued for nations through economic success, such as gross domestic product per capita (GDP) and public health success, and life expectancy or infant mortality. Subjectively, well-being can be measured by the quality of life indicator (QOL) developed by the WHO, which measures "individuals' perceptions of their position in life in the context of the culture and value systems in which they live and in relation to their goals, expectations, standards and concerns" [WHO 98]. This quality of life indicator is built on a hundred variables that explore different aspects of life. This indicator has been validated for several countries and in different contexts for culture, health, well-being and occupational stress [KUY 94, WEA 04].

Another subjective approach to measuring well-being is to question people on their level of satisfaction or dissatisfaction with life. The scores reported in the World Happiness Database (WHD) are based on answers to a simple question: "How satisfied or dissatisfied are you with your life?" [VEE 93, VEE 04].

If we consider well-being to be multidimensional, we then need to use several dimensions. Three objective measures of well-being (economic level, economic inequality and healthy life expectancy) and a subjective measure of well-being (the World Happiness index) show good mutual correlations (Figure 8.1).

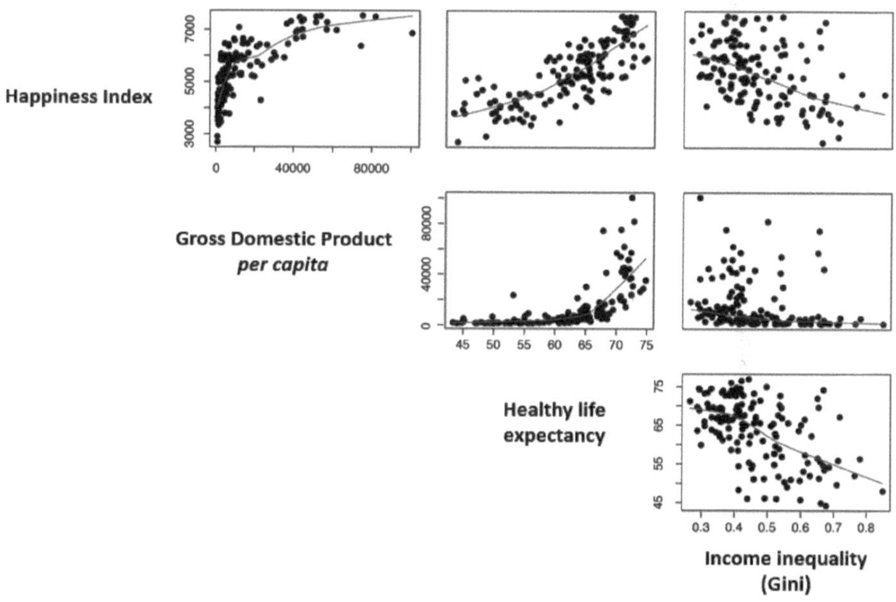

Figure 8.1. *Relationship between the Happiness Index from the World Happiness Database and three objective measures of well-being: economic level as measured by GDP per capita (in USD, World Bank data), healthy life expectancy (in years, WHO data) and the Gini Income Inequality Index (World Bank data) (all of these relationships are statistically significant). For a color version of the figure, see www.iste.co.uk/morand/biodiversity.zip*

Countries with a rich economy, long healthy life expectancy and low socio-economic inequalities have a high level of subjective happiness. As noted by McGillivray [MCG 07], the importance of using multiple measures for happiness is significant. Indeed, even if they are correlated, like in Figure 8.1, they can help us to grasp the essence of different concepts of human well-being. Thus, the happiness index increases linearly with healthy life expectancy, whereas the latter actually only increases from a certain level of wealth. The happiness index no longer increases from a certain level of wealth (measured by GDP per capita). Easterlin [EAS 01] examined the links between income and happiness, and found that individuals with higher incomes were happier than those with lower incomes. However, when a given cohort of people was tracked over time, the levels of happiness remain constant even with a substantial increase in income. This is the "Easterlin paradox", which we can see in Figure 8.1.

The subjective and objective indicators of well-being that are measured at the country level do not appear to be correlated with the state of biological diversity in the country (as measured by the number of mammalian or bird species). However, epidemiological studies have shown positive correlations between the life expectancy of people and their access to green spaces [TAK 02]. Can individual well-being and biodiversity be linked and if so, how?

8.3. Psychology and the natural environment

According to a theory by American psychologist W. James at the end of the 19th Century (quoted by [KAP 95]), all individuals possess both direct mental attention and deliberate and voluntary mental attention for aspects considered to be important to themselves. Voluntary mental attention allows the brain to recover (or restore itself) before returning to direct mental attention. According to Kaplan [KAP 95], natural environments provide the best conditions for mental recovery before moving to direct mental attention.

Geographer Roger Ulrich developed an affective theory that considers that natural parameters, such as landscapes, can produce emotional states of well-being that are detectable by psychological and neurophysiological measures. Ulrich proposed that immediate (unconsciously triggered) reactions as well as emotional responses play a central role in physiological and behavioral responses [ULR 91]. Studies have thus shown a positive impact of visual contact with nature, especially for hospitalized patients who present improved states when they see natural environment landscapes [ULR 84]. Similarly, the psychophysiological recovery of individuals under stress is better when they are exposed to natural environments [ULR 83].

Stephen and Rachel Kaplan developed a cognitive approach which assumes that the perceptual process in humans consists of extracting information from the environment, suggesting that humans seek meaning in their environment in order to engage in it and explore [KAP 89]. They identified four predictor variables. Two of these, coherence and readability, make it possible to understand the environment, and two others, complexity and mystery, encourage us to explore it. Coherence is the ability to organize or cognitively understand a visual scene. A visual scene comprising about five main units will be perceived as coherent. A large number of elements or lack of brightness will be considered as lacking coherence. Readability is the ability to predict and maintain orientation as we move through a visual scene, which implies a sense of security. Readability is similar to Appleton's prospect-refuge theory. Readable scenes are easy to monitor, unlike unreadable scenes. Complexity (also known as variety or diversity) is a component that characterizes the viewer's involvement and ability to keep his or her attention on a visual scene without creating boredom. More complex scenes tend to be preferred to simpler scenes. Finally, mystery (or anticipation) is the promise that more information will be deduced from the visual scene by exploring it.

The concept of Affinity towards Diversity is defined as the individual's predisposition to appreciate the dynamic interaction between humans and nature in everyday situations [DUA 12]. Individuals who project themselves into the future and possess socialized behaviors, such as altruism and cooperation, tend to adopt a high pro-social orientation, which translates into pro-environmental behaviors. This attitude goes beyond passive acceptance or tolerance, but includes an emotional component that expresses preference for nature, a sense of guilt for the deterioration of natural resources, and discomfort for actions undertaken by individuals or companies that affect the environment [COR 09].

8.4. Evolutionary psychology and well-being

Studies in the field of evolutionary biology have focused on the interactions between cultural evolution and human biological evolution [SEY 16], with physiological adaptations to products from plant and animal domestication, such as lactose tolerance [LAL 10] or edible plants. An example of cultural adaptation involves increasing the diversity of meals that contain spices with anti-bacterial properties when moving from high latitudes to tropical regions. Diversity in spicy dishes is linked to increased risk of food poisoning from bacteria for which development is favored by rising temperatures and also from increasing parasite diversity (Box 8.1).

This perspective of coevolution between human adaptation and evolution of the environment was put forward by Gual and Norgaard [GUA 10], who suggested integrating the interactions between three systems: biophysics, biotic and cultural.

Pharmacopoeia and the cultural transmission of plant use for medicinal and therapeutic purposes have reached decisive and unique importance for humans (Chapter 7). The use of aromatic plants and spices in the preparation of food is shared by all cultures. Some culinary plants such as garlic, onion and also chili pepper possess anti-bacterial properties and inhibit the growth of pathogenic bacteria in food [BIL 98]. However, these plants offer little nutritional value and require significant agricultural labor time, potentially to the detriment of more energy-efficient crops. The advantage of heavy use of spices should be sought in ecological and geographical conditions that favor a high prevalence of bacterial or parasitic diseases of food origin. Such a relationship between prevalence of pathogens and use of spices was tested using a corpus of 4500 traditional kitchen recipes from dozens of countries [BIL 98]. A significant correlation can be seen between the average annual temperature and the percentage of recipes that include spices. In the tropical regions, which are most subjected to heat-induced food infections, people traditionally use more culinary spices to cope. Such a correlation can also be observed within countries, for example, between the North and South of the United States, or between Northern and Southern China.

Box 8.1. *Spicy dishes and parasitic pressure*

Evolutionary psychology is a relatively new area of research that studies the origins of psychological adaptations of humans in response to social and ecological environments [BOL 11]. The origin lies in the early stages of modern human evolution. As such, seasonal depression may have an evolutionary origin. It appears to increase in prevalence from low latitudes to high latitudes [YAN 10], which could be the consequence of a long human evolution in inter-tropical environments with little annual variability in terms of photoperiod. The increase in seasonal depression syndromes is thus an expression of poor adaptation to seasonal environments, which is a feature of high latitudes.

According to hypotheses in evolutionary psychology, the evolutionary origin of human behavior is to be found in the natural environments that constituted the life of all humans until very recently. Zelinski and Nisbet [ZEL 14] suggested that humans have innate needs (in other words, natural selection) to associate with certain other life forms because of adaptations to the natural environment that have given them the capacity to establish cognitive and affective affiliations with components of these environments. Evolutionary psychology predicts that observing current behavioral patterns can provide an insight into their evolutionary origins. Thus, Tooby and Cosmides [TOO 90] suggested that adaptation in ancestral environments explains the current emotional responses of humans. These adaptations provided selective advantages for ancestral human populations, such as certain aspects of fear towards certain animals or biophobia. The hypothesis of biophobia, which is derived from evolutionary psychology, assumes that certain aspects of biodiversity generate fear and avoidance because of their association with danger [ULR 93]. The best-known example is fear of snakes (Box 8.2).

Mammals have brain structures that are involved in alertness, fear, learning and memory associated with fear stimuli such as those related to predators. Some of these areas that were related to the visual system developed in primates and are linked to visual systems. Isbell [ISB 06] showed that the variability of brain expansion linked to visual systems in primates is related to predation pressure by snakes. This variation in primate brains coincides with evolutionary coexistence variation with venomous snakes: the Malagasy Prosimians never coexisted with venomous snakes, the Platyrrhinian monkeys of the New World experienced an interruption of this coexistence and the Catarrhinian monkeys of the New World coexisted continually with poisonous snakes. Souchet and Aubret [SOU 16] explored the aversion of humans to snakes and showed that the origin of this behavior must be sought in response to an aposematic signal, i.e. stemming from an adaptive strategy to the emission of a warning signal. In this case, it was a triangular shape signifying a tooth, a spike or a claw. According to Souchet and Aubret [SOU 16], the evolution towards the triangular head of snakes is more of an adaptation to anti-predator defense, resulting from this aposematic signal. That would then be a real coevolution between humans (brain development and biophobia for snakes) and snakes (morphological evolution).

Box 8.2. *Biophobia of snakes*

8.5. Theories of habitat and visual refuge, topophilia and biophilia

The cognitive and affective links between humans and their environment have been explored by various ecological and evolutionary theories. The most important of these are: habitat theory, prospect-refuge theory, topophilia and biophilia.

Habitat theory was one of the first to examine the relationship between the original habitat during evolution and adaptation of early humans in Africa and its consequences on the behavioral responses of contemporary humans [ULR 77]. According to Orians [ORI 80], savanna-like environments with trees scattered in a grassland matrix should be the preferred environments for contemporary humans, eliciting positive emotions. Thus, our preference for open urban parks with scattered groves is an emotional consequence of the evolution of early humans in open savannas [BAL 82].

The prospect-refuge theory was proposed by the English geographer Jay Appleton. In his book "The Experience of Landscape" [APP 75], he described landscape as visual perspectives of refuges and dangers. This theory, which is also based on an ancestral adaptation of human populations, predicts a preference of choice for places that allow us to see and stay out of sight of potential predators. The theory predicts that humans are attracted by panoramic views and places that provide visibility of the surroundings and refuge (such as groves and caves). This theory explains human preferences for panoramic views and certain aspects of aesthetics [RUS 03].

Geographer Yi-Fu Tuan [GEO 74] coined the term topophilia as the study of human perceptions, attitudes and values in a local environment. According to Tuan [TUA 74], the feeling of belonging to a place is manifested by a universal affective bond that is the characteristic of basic human needs. Topophilia is the link between humans and the non-human world, which can be explained by both cultural learning and innate cognitive behaviors that arise from human evolution. One of the implications of topophilia is that humans have the ability to learn to like what becomes familiar, which is the love of locality [ORR 93].

The evolutionist Edward O. Wilson [WIL 84] proposed the hypothesis of biophilia, which suggests that nature is indispensable to the psychological and social equilibrium of humanity. Biophilia assumes that contact with nature is fundamental to psychological well-being and personal development [KEL 93]. It is a process of human evolution that integrates psychological and emotional aspects. Many components of our behaviors find their adaptive origins in the millennia of human evolution, and the study of these components, such as evolutionary psychology and evolutionary medicine, explains the links between biodiversity, health and well-being. As Seymour [SEY 16] pointed out, our historical memory remembers the beauty of the gardens of Babylon, which would be credited to the existence of biophilia among the first urban dwellers (gardens with identical functions were also created in the cities of ancient China). Biophilia hypotheses also explain the evolutionary origin of our aesthetic perceptions [RUS 03].

The prospect-refuge, habitat and topophilia theories are all linked to the fundamental concept of biophilia that concerns people's attachment to nature [KAH 97]. This attachment is achieved through the pleasure caused by the shape and spatial arrangement of elements in a landscape. Environmental perceptions and preferences result from the adaptive need to survive, which now manifests itself through well-being [WIN 11]. Howell *et al.* [HOW 11] suggested that richness and diversity of nature have favored serenity and attention to human feelings. Although biophilia explains the existence of a human perception that is intrinsic to nature, topophilia explains the attachment to local environment from the point of view of individual and collective development. In some ways, topophilia is more closely related to environmental psychology and traditional ecological knowledge.

The cognitive perceptions of nature would therefore originate in the evolutionary history of the human species, and maybe great apes [VER 02], in coevolution with other species. As for the representations of nature, which appear to be so different between human societies, they would come from a coevolution of societies with their environments.

8.6. Implications and applications of biophilia

Research on evolutionary psychology and biophilia has applications in medicine, architecture and urbanism. These applications aim to create living environments that promote good social interactions, improve individual well-being and health, and lower the risk of developing psychological illnesses such as depression [GUL 00]. They assume that connectivity to nature generates positive effects on health [BIR 07].

Various studies have shown that active and passive contact with gardens offers psychological, emotional and social benefits. Cooper-Marcus and Barnes [COO 95] showed that doing gardening improves the healing of hospitalized patients. People with dementia or who have had a stroke show improved mobility and dexterity, as well as better social skills through gardening activities [RIV 13].

A national US-wide cohort study on nurses' health in 2000 and 2008 showed a significant decrease in mortality of women who live in green areas, estimated by satellite imagery with NDVI vegetation index [JAM 16].

Another study on health data in 15 US states from 1990 to 2007 measured the effects of deforestation induced by a forest pest [DON 13]. Mortality from cardiovascular and respiratory diseases is correlated with loss of trees due to the impact of this forest disease. Degradation of the natural environment subsequent to the emergence of a disease that affects forests in turn affects human health.

The concept of a "therapeutic landscape" is an approach to health promotion following the Ottawa Charter [WHO 86]. William Gesler introduced the concept of therapeutic landscape in his book "The Cultural Geography of Health Care" [GES 91]. He introduced it as a geographical metaphor to help understand healing processes related to local situations and settings [WIL 99, WIL 10].

The origins of urban planning include health and green spaces and date from Rauch [MAR 09a] in his report on "Public parks: their effects on the moral, physical and sanitary conditions of the inhabitants of large cities. With a special reference to the city of Chicago" (quoted by [MAR 09a]). This report led the city of Chicago to develop an urban planning that included many parks and green spaces. Many cities have implemented biodiversity action plans based on the concepts of biophilia [BEA 11] with an integration of nature in town planning and architecture. Thus, some cities have introduced provisions in their municipal law that encourages green spaces and biophilic design.

The presence of green spaces also has an influence on cognitive development. A study conducted in Barcelona assessed the association between exposure to green spaces and the cognitive development of children in 36 primary schools. Working memory was higher among children who were in contact with green spaces at school, around the house, or on the school–home route [DAV 15].

8.7. Traditional knowledge and well-being

Biocultural diversity includes both biodiversity and cultural diversity (Chapter 2). According to Maffi [MAF 08], biocultural diversity is an awareness of the interactions that have developed over time between humans and their environment on a local level. The notion of "traditional ecological knowledge" (TEK) allows us to see how nature "forms, penetrates or even permeates human cultural expression and vice versa" [MAF 08]. Some links between cultural diversity and biodiversity may have coevolved (Box 8.3).

Nichter [NIC 08] defined the term perception as the process of collecting information through all our senses, followed by the organization of that information to give it meaning. Health and illness are therefore perceived cognitively and viscerally. Nichter [NIC 08] also told us that "understanding ethnophysiology requires thinking ecologically". An example is the perception of synergistic relationships with intestinal parasitic worms. In many countries, communities consider that a limited number of parasitic worms living in the body play an important role in digestion. Thus, in a village in Kenya, people consider these parasites as positive agents for digestion, with which we must maintain a balanced relationship [GEI 98]. The number of worms is only controlled when they become too numerous. However, people are not in favor of eliminating them because worms are perceived to fulfill essential functions in the body. Ethnophysiology analyzes perceptions of the body and parasites to find a link with knowledge of the evolutionary biology of parasitic nematodes, which are now seen as important elements of immune homeostasis (see Chapter 4).

Box 8.3. *Perceptions of health and ethnophysiology*

The positive correlations between biodiversity and linguistic diversity per country (Figure 2.3) and the risks of extinction of biodiversity and linguistic diversity (Figure 3.6) demonstrate causal relationships between cultures and biodiversity, according to Maffi & Woodley [MAF 10]. They pointed out that “in the same way that a healthy planet requires biological diversity, a healthy cultural world requires linguistic diversity”. Loss of language threatens the expression of a population and therefore a fundamental human right. Language rights were denied for at least 400 years with the colonization and destruction of native communities, the development of European nation-states and the eradication of local languages and patois. This also occurred in some post-colonial states.

Ethnobiologists and environmental anthropologists have documented traditional ecological knowledge (TEK) and emphasized the value of traditional ecological knowledge in the welfare and livelihoods of local communities, in the sustainable use of resources, conservation of the environment and analysis and monitoring of long-term ecological changes [MAF 10]. Changes in local ecosystems have profound implications for both knowledge and biodiversity. Such changes include the loss of cultural identity and acculturation to a dominant way of life, the shift from local language to majority language, integration into a market economy, and the loss of local decision capacity and self-sufficiency. According to a survey conducted in Canada, 63 % of First Nation respondents identified loss of land and cultural knowledge as a deteriorating health condition [UNI 06]. This was supported by another study which demonstrated that health and well-being were higher in First Nation communities that had retained their language and culture or who had taken steps to rebuild the cultural continuity of their lands [CHA 98].

8.8. Conclusion

Finally, one of the proofs of our affective and emotional ties with life is given by studies on the stress levels of staff in abattoirs [DIL 08] or on people who are responsible for widespread slaughter of animals during infectious disease epidemics, such as foot-and-mouth disease in England [MOR 08a] or more recently, avian influenza. Through their activities that are so closely linked to animal suffering, veterinarians are among the occupations most affected by suicide [BAR 08].

As highlighted by Van Leeuwen *et al.* [VAN 99] and, more recently, by Seymour [SEY 16], few studies have incorporated socio-economic and biophysical environments with relevant ecosystem characteristics (diversity, interactions, habitat and bioclimatic characteristics) and different dimensions of human health and well-being [REE 96]. More recently, the "One Health" concept has been launched as a disciplinary and holistic integration [ZIN 12] which aims to achieve optimal health at the human–animal–environment interfaces at local and national levels. In addition to criticisms about the emphasis on infectious diseases and the lack of social perspectives [WOL 14], it is regrettable that little consideration is given to cultural and environmental coevolution and traditional ecological knowledge.

Further development in this area therefore requires taking into account the links between biodiversity and human well-being [NAE 16], and ecosystem services are one of the solutions presented to meet this challenge.

9

Ecosystem Services for Health and Biodiversity

9.1. Introduction

The planet has entered a new geological era: the Anthropocene [STE 07]. Through its relationship with the planet, the human species has demonstrated its ability to modify major geochemical cycles, the climate and the biosphere [MCN 00, WAT 16]. Human health has never been better than today, on average, thanks to improvements in public health that are associated with an increase in the wealth of nations. However, the corollary of this economic performance lies in its environmental impacts [WHI 15]. The degradation in function of ecosystems because of agricultural intensification and widespread use of various biocides is cited as a factor that aggravates health risks (Chapter 5). Changes and loss of biodiversity affect the epidemiological environment through the emergence or re-emergence of new infectious diseases (Chapter 3) and also non-communicable diseases (Chapter 4). Most ecosystems are considered to be more or less degraded [MA 05], which increases insecurity and conflict. Human well-being is clearly dependent on good social relationships that are reflected in cognitive and affective ties to the natural habitat and the environment (Chapter 8). There clearly exists a link between ecosystem health, well-being and human health [MA 05].

The notion of "ecosystem service" emerged from the ecosystem approach promoted by the Convention on Biological Diversity (see Chapters 1 and 11). Ecosystem services were defined by Costanza *et al.* [COS 97a] as "the benefits that people get from ecosystems". Daily [DAI 97] provided a more in-depth definition of ecosystem services as "the conditions and processes through which natural ecosystems (and the species that make them up) sustain and fulfill human life". Ecosystem services are therefore defined relative to human needs and are assessed according to their socio-economic and ecological values. The ecological functioning of natural systems is thus analyzed in terms of social utilitarianism.

According to the Millennium Ecosystem Assessment [MA 05], ecosystem services that are required for human health and well-being include provision, regulatory and cultural services (Chapter 1, Figure 1.1). Nearly 10 years after the formalization of the notion of ecosystem service, the Intergovernmental Science-Policy Platform on Biodiversity and Ecosystem Services (IPBES, Figure 1.2), emerged as an international institution that mobilized international scientific expertise in biodiversity and ecosystem services (see Chapter 11). Its objective was to inform decision-making and public policies in a context where the ecosystem approach and ecosystem services were considered as the preferred tools of international expertise [MOR 11a].

In the previous chapters, we explored the negative aspects of biodiversity change and the beneficial effects of environmental and landscape elements on human health and well-being. However, it appears that the results from studies are still far from proving a link between the level of biodiversity (ecosystems, habitats and species) and human health. In this chapter, we discuss how the concept of ecosystem health has led to the concept of ecosystem services and how this concept responds to research on the links between biodiversity and health.

9.2. Environmental impacts and well-being

In the previous chapters, we explored the links between biodiversity, economic development, health and well-being (see Figures 3.2 and 8.1), but these do not show any association between loss of biodiversity, as estimated by the number of endangered mammals and birds, and human well-being and health. However, the effects of environmental impacts on ecosystems are not limited to loss of animal or plant species and must take other factors into account such as land use changes and environmental pollution. Bradshaw *et al.* [BRA 10] developed a proportional composite environmental index per country. This index is calculated from loss of forest habitat, transformation of natural habitat (agriculture or urbanization), fisheries (impact on living resources), use of agricultural inputs, water pollution, proportion of endangered species and emission of greenhouse gases (CO_2). Low values of this environmental impact index suggest a high level of actual impact on the environment [BRA 10]. We can show that this environmental impact index is negatively correlated with two indices for well-being at the country level: the objective index of healthy life expectancy and the subjective index of happiness (Figure 9.1, see also Figure 8.1). Degradation of the environment, which is reflected by low environmental impact index values, is correlated with poor objective and subjective values of health and well-being.

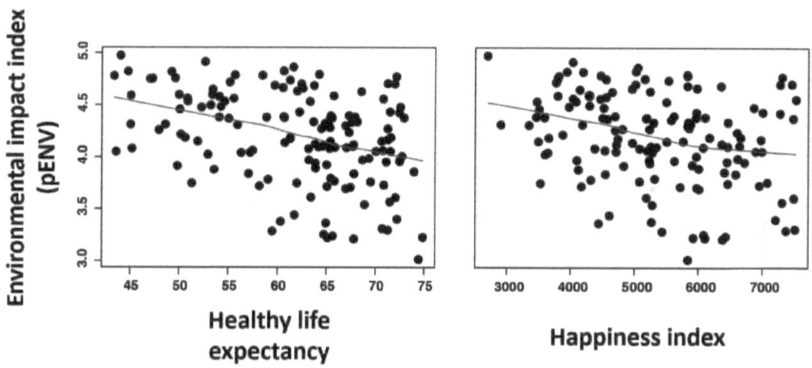

Figure 9.1. *Correlation between impact on the environment and human well-being at the country level. The environmental impact is estimated by the proportional composite environmental index, which incorporates loss of forest habitat, natural habitat transformation, fisheries, use of agricultural inputs, water pollution, proportion of endangered species and greenhouse gas emissions (CO_2). Well-being is valued by healthy life expectancy (in years, WHO data) and the subjective happiness index (World Happiness Database). There is a negative correlation between the environmental impact index and, on the one hand, healthy life expectancy ($P < 0.001$), and on the other hand, the happiness index ($P < 0.001$). For a color version of the figure, see www.iste.co.uk/morand/biodiversity.zip*

One of the objectives of the concept of ecosystem health was to link ecosystem degradation with human health and well-being [COS 99a].

9.3. Health of ecosystems

The concept of ecosystem services is further elaborated through ecosystem health [RAP 98]. Characterizing ecosystem health makes it possible to infer conservation applications in terms of natural capital and human health [KAR 11]. The idea of assessing ecosystem health (Ecosystem Health Assessment) in the management of the environment emerged in the late 1980s [JØR 10]. Based on the observation that an ecosystem is not "healthy", a diagnosis is made to find the causes and recommendations to restore the ecosystem to its "normal" function [SCO 00]. The initial diagnosis and final mediation processes require functional ecological indicators to be developed [JØR 10]. The concept of ecosystem health has fostered the development of ecological engineering and ecological restoration [ARO 07].

An ecosystem is considered to be "healthy" if: it is not degraded, it can maintain its organization and autonomy over time, and it is resistant to stress [RAP 98]. According to Costanza and Mageau [COS 99a], an ecosystem is healthy and sustainable if it has the "ability to maintain its structure (organization) and function (vigor) over time in the face of external stress (resilience)". The structure of an ecosystem is defined by biological diversity indices and ecosystem function indices such as those developed by Ulanowicz [ULA 86]. Vigor can be estimated by primary production, referring to metabolism for a biological organism, or to the gross national product in reference to the economy [COS 99a]. Resilience is defined either as the time it takes for a system to recover from stress [PIM 84], in other words the time needed to return to the initial state, or as a stress value beyond which the system cannot restore its initial state [HOL 86, GUN 09].

A "non-healthy", degraded and poorly resilient ecosystem is implicitly considered to have negative effects on the health and well-being of those who live or depend on it. However, despite there being a broad scope of literature on ecosystem health and its indicators, studies that directly link health of ecosystems to human health are rare [REP 98, RAP 03, JØR 10] if we exclude industrial pollution of ecosystems and their direct impact on human health.

Some criticisms of this concept ought to be highlighted. A first criticism concerns the need to establish a state of reference. Indeed, the ecological health of an ecosystem is evaluated with reference to habitats that are considered as little or not at all degraded and are characterized by their integrity and vigor (state of good ecosystem function). A second criticism concerns the use of the term health, which is deliberately taken as an analogy of human health, to analyze the function or dysfunction of ecosystems [CAL 92]. The health of ecosystems leads us to think of the ecosystem and the entire planet as a living entity that must maintain its homeostasis (Gaia hypothesis, [LOV 95]). Finally, some criticism focuses on the metaphor of the "ecosystem as an organism" [RAP 98] or based on ecological theories that advocate balance, whereas animal and plant communities are dynamic assemblages of species [DEL 97, ROH 05] that are governed by coevolving processes [LEV 92].

The concept of ecosystem health is closely tied to social values. As in the case of human health, the health of ecosystems integrates multiple factors: ecological, social, economic and political [RAP 95]. The concept of ecosystem health has directly contributed to the development of the concept of ecosystem services: healthy ecosystems have the capacity to provide a full range of ecosystem services [COS 97b, LU 03, MYE 13], while ecosystems that suffer from ecological stress would reduce the quality and quantity of ecological services [CAI 95]. Altering ecological functions would affect the health of ecosystems, as well as the ecosystem services provided by ecosystems and ultimately health and human well-being [RAB 13]. The concept of ecosystem health is linked to the notion of natural capital [COS 97a] (Box 9.1), making it possible to economically evaluate ecosystem services (TEEB, Bishop [TEE 12]).

Sustainability can be defined as the amount of consumption that can be maintained indefinitely without degrading capital stocks, including natural capital stocks. Traditionally, capital is defined as the production of a means of production. The term natural capital is based on a more functional definition of capital as "an asset that produces a flow of goods or valuable services in the future". The functional relationship of natural capital is therefore a stock of natural products that produces a flow. A stock or population of trees or fish provides an annual flow or yield of new trees and fish. This flow can be sustainable over years. A sustainable flow is natural income, and stock that produces a sustainable flow is natural capital. Natural capital can also provide services such as waste recycling or water abstraction and erosion control, which are also counted as natural revenues. Natural capital, ecological capital, is the structure and functions of the ecosystem that support the creation and flow of natural goods and services that are valuable to humans.

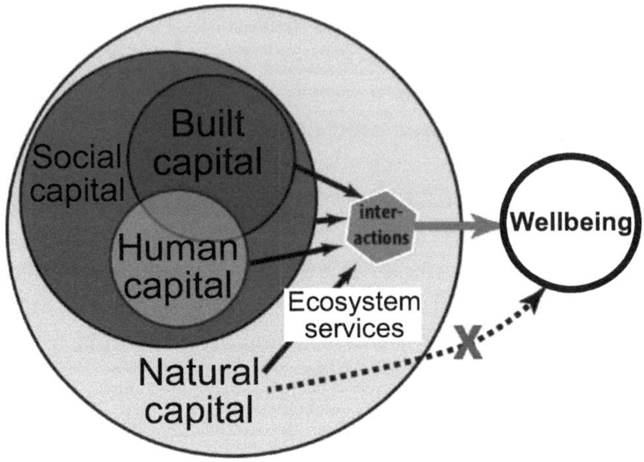

Box 9.1. *Natural capital and ecosystem services (according to Costanza in [JØR 10, RUH 07]). For a color version of the figure, see www.iste.co.uk/morand/biodiversity.zip*

9.4. Ecosystem services

The term "ecosystem service" refers to the delivery, provision, protection or maintenance of goods and benefits that humans derive from the ecosystem (Chapter 1, Figure 1.1). In a non-exhaustive list, De Groot *et al.* [DEG 02] identified at least 32 biological, physical, aesthetic, recreational and cultural ecosystem services.

Wallace [WAL 07] proposed a conceptual framework for building services based on the identification of human values. Within these human values, protection against diseases, obtaining adequate resources (including food) and socio-cultural satisfaction (preservation of biological and cultural diversity) are the values that link biodiversity, health and well-being (Table 9.1). Individual and collective ecosystem services can then be identified, as well as ecological and socio-economic mechanisms, capital (social, economic, political) and assets (ecological, socio-cultural), which must be managed to produce these services.

The relevance of such a conceptual framework lies in the fact that from the outset, it imposes scientific pluridisciplinarity (ecologist, agronomist, anthropologist, political scientist, economist, etc.) and plurality of points of view (scientific, political, layman).

Human values	Ecosystem services at the individual and collective level	Mechanisms and capital/assets need to be managed to produce ecosystem services
Protection against diseases/parasites	Regulation/protection against diseases and parasites	Biological regulation
Adequate resources	Production (food)	Pollination
Sociocultural	Opportunity values	
Capacities of cultural and biological evolution	Socio-cultural interactions	

Table 9.1. *Human values determine ecosystem services and the mechanisms and capital to be managed to produce these services (according to [WAL 07])*

Wallace [WAL 07] then proposed a functionalist approach to services, starting from an overview of the functioning of ecosystems in terms of energy where abiotic elements (climate, temperature, precipitation) are the determinants of biological productivity (plant and animal biomass). Modes of land appropriation and land use explain the notion of landscape and sharing between animal and plant communities (domestic and wild), predators and parasites (Figure 9.2). Identified biological regulations, host–parasite interactions, pollination, predation and competition all contribute to the production of ecosystem services, which we here limit to diet, disease regulation and conservation.

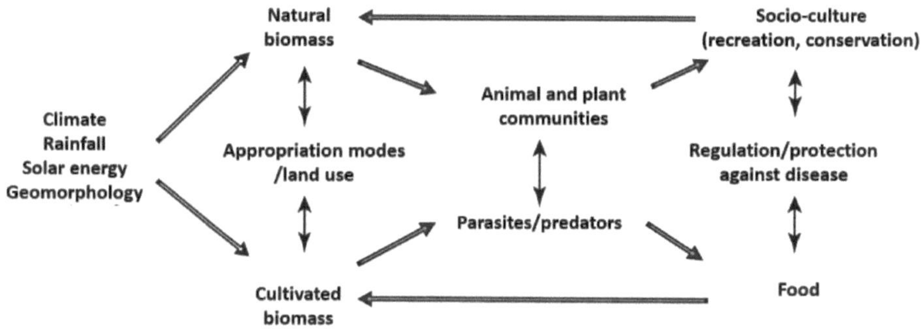

Figure 9.2. *Diagram of the links between ecological processes that lead to three ecosystem services (inspired by [WAL 07], see [PED 15])*

9.5. Ecosystem services and health

The literature on ecosystem services is substantial, but only a small number of studies directly analyze the links between ecosystem services, particularly for regulation and health [SAL 09, LÉV 12a], compared to studies carried out on other services such as climate control or water purification. It is also worth noting the low number of indicators that has been developed and validated for all regulatory services [WAL 09]. Studies have been carried out to ascertain the extent to which biodiversity can provide an ecosystem service for regulating diseases for humans, animals or plants. These have yielded contradictory results [CAR 12], which can be explained through the different methodologies used.

In a literature review on the links between ecosystem services and human well-being, Martinez-Juarez *et al.* [MAR 15] also noted that studies on the topic were few and far between. Most studies showed positive effects of the environment. However, these effects were not always statistically significant and they often interacted with other factors, such as socio-economic factors. Particularly for urban ecosystems, the health and well-being benefits that humans derive from contact with elements of nature are most significant. This was also confirmed by Butler and Olouch-Kosura [BUT 06].

The few studies that have explicitly addressed ecosystem services and health or well-being show that this service does not necessarily depend on biodiversity as such, but on the presence of biodiversity elements, such as a pretty landscape or the presence of a garden in a city. Often, the ecosystem service for health is considered as the good health of an ecosystem, ensuring proper functioning of drinking water treatment [NAS 05].

In a study that links ecosystem service and disease control, Lamb *et al.* [LAM 17] showed that seagrass beds provide purification functions for pathogenic microbes. Their presence reduces coral reef diseases and therefore they are assumed to provide an ecosystem service for the regulation of infectious diseases for important elements of ecosystems, corals, and also for humans. This study specifically focused on a component of biodiversity that plays a purification role, but the implications of this study are important in ecological engineering and restoration for beneficial effects on the health of marine organisms and humans.

A study by Hanski *et al.* [HAN 12] is one of the few studies that showed a causal link of ecosystem service for health (Chapter 4). Maintaining diversified landscapes that are rich in various pollens limits the appearance of atopic symptoms. Biodiversity and ecosystem services for health can be analyzed through immunity. Diseases associated with faulty immunoregulation and unregulated inflammatory responses are increasing in developed countries (Figure 4.1). This trend is attributable to a loss of exposure to parasite and microbial organisms (hygiene hypothesis, Figure 4.4) in human evolutionary history, which plays a role in the immune system's education and in the regulation mechanisms of immune responses. The modern urban environment has significantly contributed to the elimination of parasites from the digestive tract, ectoparasites and the reduction of human microbiota (accelerated reduction with widespread use of antibiotics, see Chapters 4 and 5), which increases our dependence on microbiota from the natural environment [ROO 13]. The causal link between microbiota diversity, habitat diversity and proper functioning of the immune system complicates the assumption made by Costanza *et al.* [COS 14] on the lack of direct movement of natural capital towards human well-being (Box 9.1).

9.6. Ecosystem disservices and health

Conservation of biodiversity can lead to health risks because of the positive relationship between biodiversity and diversity of infectious diseases at the country level (Chapter 2). Thus, a quick look at Figure 2.2 might suggest that reducing a country's parasitic and infectious burden could be easily achieved by reducing diversity in mammals and birds. However, this can be proven; biodiversity loss can be beneficial, as suggested in a study by Valle and Clark [VAL 13] which showed that the fight against deforestation in Brazil can increase the risk of malaria. The complex links between biodiversity and health caused Lafferty and Wood [LAF 13] to suggest that considering biodiversity as a protection against health risks is unrealistic. Criticisms of the dilution effect focus on this aspect. In addition, arguing that biodiversity has health benefits may be even counterproductive to biodiversity conservation objectives.

The concept of ecological services is presented as a response to the utilitarian and positive view of ecosystems, habitats and species [DUN 10a]. Based on the principle that nature can kill (diseases, predation, snake bites, etc.), it would seem wise to manage species and ecosystems to minimize ecosystem services. The benefits and advantages of change in ecosystems and habitats need to be analyzed [DUN 10b]. Ecosystem dysfunctions are ecosystem functions or properties that cause negative effects on human well-being and are perceived as harmful, unpleasant or undesirable [ESC 11] (Table 9.2).

Ecosystem disservice	Cost
Human damages from diseases, bites from insects or other wild animals, reservoirs of vectoral diseases, allergies	Financial: public health, disease control, vector and reservoir control, irrigation, planting, pruning
Attraction of wild animal droppings, attacks on domestic animals, bites	Social nuisances
Quantity and quality of water – influx of fertilizers and pesticides	Environmental
Air pollution due to maintenance: emission of carbon dioxide and methane due to decomposition, air pollutants	
Emission of volatile organic compounds	
Introduction of invasive species	

Table 9.2. *Example of urban health-related ecosystem disservices and costs (according to [RAN 12, ESC 11])*

The concept of ecosystem disservices has been developed in agriculture to characterize the range of agroecosystem components that can reduce productivity or increase production costs. For example, the effects of diseases and crop predators or competition for water and soil nutrients by weeds [ZHA 07].

The notion of ecosystem disservices has extended to urban spaces [LYY 08]. Health issues play a significant role, such as allergy to pollen or animals (rats, pigeons, cats), as well as infectious diseases that are linked to commensal urban fauna or wildlife in urban parks [GAU 14].

The importance of values and perceptions that are associated with nature and ecosystems is also central to the notion of ecosystem services. These values may be biophobic (see Chapter 8) or even from the evolutionary psychology of disgust [CUR 01, KEL 11].

Lyytimäki [LYY 14] analyzed negative representations of nature in newspapers. These mostly concerned fears of wild animals (foxes), emerging zoonotic diseases and their vectors (ticks) or reservoirs (birds, rodents), as well as harm caused by wandering wild animals (squirrels) and inconveniences (pigeon droppings).

Criticisms of ecosystem disservices, not to mention criticisms of the very notion of ecosystem services [MAR 14], can fall into several categories. First is the negative perception of species: should they be analyzed in terms of disservices produced by ecosystems or psychological responses related to biophobia or disgust? The second is related to the state of the ecosystems to which it refers. Thus, for infectious diseases, the chapters in this book have differentiated between the link of diversity of infectious diseases/ biodiversity, and the link of infectious disease epidemics/changes in biodiversity and habitats. Ultimately, the most important criticism is the lack of an integrated vision of ecosystem services and disservices induced by practices.

Power [POW 10] presented an approach that provides a conceptual framework to analyze management practices for the potential of ecosystem disservices in agriculture (Figure 9.3), including habitat loss for biodiversity conservation, nutrient losses through runoff, greenhouse gas production, pesticide poisoning in humans and non-target species. Such an approach allows the externalities in terms of costs for health and well-being to be evaluated.

Tegtmeier and Duffy [TEG 04] estimated the external costs of agricultural production in the United States on natural resources, biodiversity, ecosystems and human health. They are valued at between \$5.7 billion and \$16.9 billion annually for agriculture, and about \$720 million annually for livestock. For human health, the cost of agriculture and livestock would be over \$1.7 billion annually. Costs also relate to pesticide-related diseases, infectious food-borne diseases and also the costs of health surveillance agencies. Further research is needed to refine these estimates and include categories that were not covered by this study. According to the authors, these costs represent such a societal burden that a restructuring of agricultural policies is necessary.

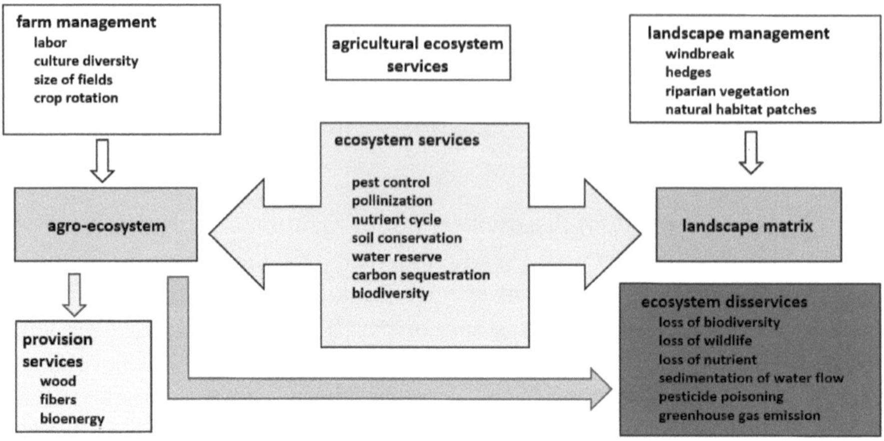

Figure 9.3. *Impacts of agricultural management on the flow of ecosystem services and disservices [POW 10]. For a color version of the figure, see www.iste.co.uk/morand/biodiversity.zip*

9.7. Compromise between services, economic development and health

The difficulty in developing and validating ecosystem services for health regulation is the evaluation of such services. Ecosystem services for disease control can be assessed through the costs of inaction such as increased morbidity or increased cost of public health services. Another approach is to identify trade-offs between ecosystem services in order to analyze the objective and subjective characteristics of health and well-being.

Barbier [BAR 11] proposed the notion of "ecological shortage" as "the result of a fundamental compromise in the use of the natural environment" (Figure 9.4). This concept highlights the beneficial effects of economic development and the beneficial effects of ecological shortages on human health and well-being. Economic development is based on the exploitation of natural resources. Their use through changes that affect land use (agriculture, urbanization, dam construction, etc.) in turn affects the quality of ecosystem services. The positive side of economic development is a general increase in wealth and consumer goods, contributing to an overall increase in human health and well-being. Exploitation of the natural environment also leads to alteration of ecosystems. This ecosystem disturbance affects the quality of various ecosystem services that contribute to human well-being, such as production services, climate and water regulation services, beneficial species' maintenance and disease control. The loss of ecological benefits and ecosystem services as a result of economic development is ecological shortage, which affects human health and well-being.

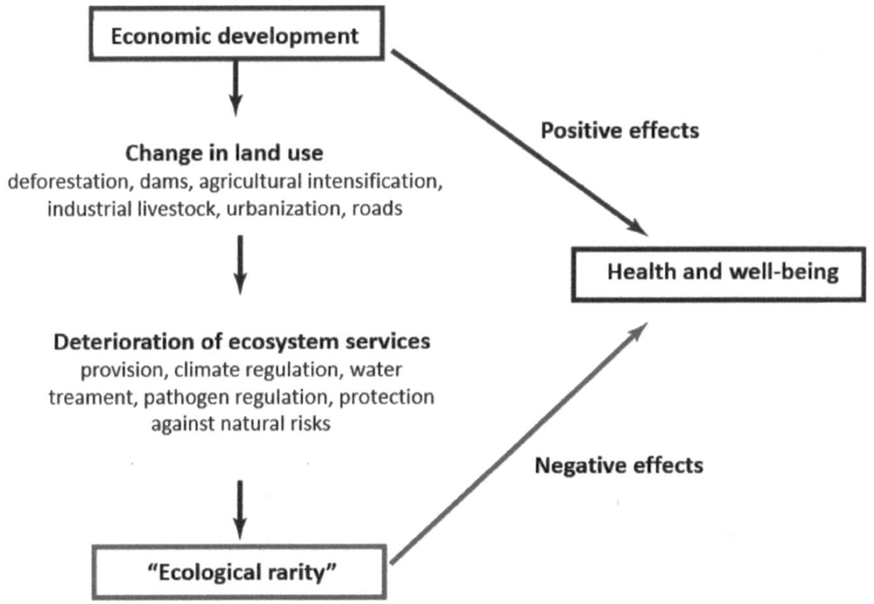

Figure 9.4. *Effects of economic development on ecosystems, with the beneficial effects of economic development on health and the negative effects of its impacts on ecosystem services and increase in ecological shortage (from [BAR 11]). For a color version of the figure, see www.iste.co.uk/morand/biodiversity.zip*

The evaluation of externalities in terms of cost on human, animal and biological diversity is a decision-making method [TEG 04].

9.8. Conclusion

In this chapter, we briefly reviewed the history of the concept of ecosystem services, which originated in the notion of ecosystem health. The concept of ecosystem services requires a dialogue between scientists, experts, politicians and citizens. Sharing knowledge implies understanding and exchanging views, values and representations of the world (academic, political, socio-cultural), as well as perceptions of time (evolutionary, ecological and seasonal) and space (global and local).

The birth of this concept in the development of ecosystem health research has led to health being integrated into planning policies. Ecological engineering or ecological restoration can thus have direct public health objectives.

It should be also noted that medicinal plants and associated knowledge on traditional pharmacopoeias, protected areas and conservation areas are also presented as ecosystem services or as providers of services for human health while preserving biodiversity [WWF 10].

The concept of ecosystem services allows the economic evaluation of natural capital, which is often presented as a means of ensuring sustainable development that respects biodiversity [CHE 09]. The concept of ecosystem services is ultimately a tool for building ecosystem management scenarios that require the evaluation of services and disservices as well as trade-offs between different services. The external costs of human health and animal health that are caused by habitat management should be integrated into agricultural policy or urban planning.

10

Biodiversity and Health Scenarios

10.1. Introduction

In "An Essay on the Principle of Population" (1798), Thomas Malthus established one of the first scenarios that linked natural constraints (agricultural production) with demography and economic growth. This essay was an essential contribution not only to the development of economic theories but also to ecological and evolutionary theories with the concept of a limiting capacity within a given environment. The Malthusian view of human demography is at the core of the book by Ehrlich and Ehrlich "The Population Bomb" [EHR 68], which called for active control of births in order to avoid famine, resource depletion and corporate collapse.

However, it was the Meadows report "The Limits to Growth" [MEA 72] that marked the beginning of global environmental scenarios that were based on modeling and computer simulations. The Meadows report was also part of a neo-Malthusian view that biological and therefore human populations tend to grow until they reach the limit that their environments can bear [BAR 11a] (Figure 10.1). The initial model used in "The Limits to Growth" takes many factors into account while addressing the interactions between the five main factors: human population, non-renewable resources (minerals), renewable resources (agriculture), capital resources and pollution. A review of the scenarios from the first Meadows model was produced in 2004: "Limits to Growth: The 30-Year Update" [MEA 06].

Negative criticism on the report by Meadows mainly focused on his Malthusian view of the dynamics of human societies and errors in his predictions, particularly on the depletion of certain non-renewable resources predicted for the end of the 20th Century. More recently, critics have been more positive, pointing out that scenarios in the Meadows report continue to provide useful projection elements, while stressing that models should now integrate climate change and the biodiversity crisis [BAR 11a].

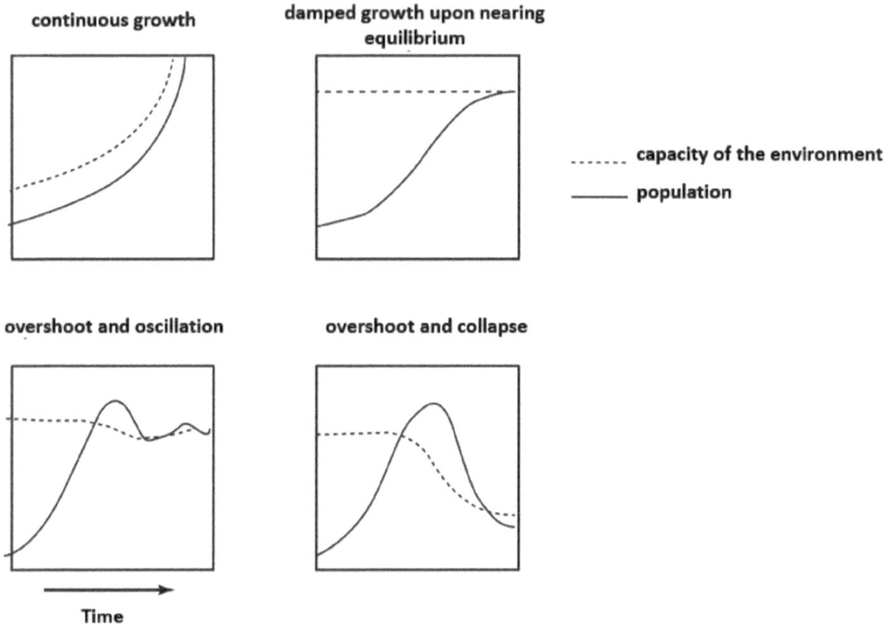

Figure 10.1. *The relationships between demographic growth patterns and limiting capacity of the environment [MEA 06]*

The Malthusian approach to interactions between human societies and their environment, with the stance that humans are disruptive agents of ecosystems, has hampered collaborations between ecological and social sciences [FIS 16]. However, thanks to the UNESCO's "Man and the Biosphere" (MAB) program, which was initiated in the 1970s (Chapter 11), collaborations were nonetheless able to be developed.

This chapter illustrates some elements of the prospects, scenarios and models for studying interactions between biodiversity and health.

10.2. Prospects and global scenarios

The elements of prospects or global scenarios (Box 10.1) are generally drawn up in broad sectoral areas: demography, urbanization, globalization, land use, agriculture and livestock, and living resources. These areas, which are usually analyzed individually but sometimes in interaction, have independent effects on health and biodiversity and on the nature of biodiversity/health links.

> Extrapolation is the simplest method of projection into the future. It is based on data from the past and assumes that trends in the past will continue into the future. Extrapolation is used to project the change in distribution of an infectious agent vector, such as mosquitoes, with climate change.
>
> Phenomenological models are used to combine quantitative components. They help to unravel the underlying mechanisms and carry out projections on the values of parameters. Phenomenological models are widely used in ecology, hence the relationship between the abundance and distribution of a species (such as a parasitic species).
>
> Prospecting was introduced in the United States by the Rand Corporation with the development of the Delphi method and the scenario method, which was based on expert opinion. The scenario method requires analyzing different possible states of the future, according to different alternative conditions. Experts construct extreme, best and worst-case scenarios, or typical and alternative scenarios [ROW 99].
>
> Futurology is the study of possible futures, both probable and preferable, using quantitative and qualitative data.
>
> As Sutherland [SUT 06] pointed out, predicting consequences linked to the emergence of new conditions is particularly difficult, especially for the emergence of a very rare event.
>
> Traditionally, projections contemplate the various current options that may influence future states of the system or domain being analyzed. An alternative method is to determine the strategies needed to reach a desired future state or to avoid paths that lead to an unwanted future state [DUP 04].

Box 10.1. *Definitions: scenario, model, prospect, projection (adapted and complemented by Sutherland [SUT 06])*

10.2.1. *Demography*

Global population growth projections are regularly reviewed by UN agencies. According to the latest United Nations projection, the world's population is expected to reach 11.2 billion by 2100 [UNI 15a] (Figure 10.2). Several important characteristics for the interactions between biodiversity and health are associated with this growth of world population and are worth noting. Thus, the population will continue to grow in tropical regions until 2100, while it will decrease in temperate regions of the world (Figure 10.2). Human population growth will be affected in areas that are rich in biodiversity and endemic infectious diseases (Chapter 2). Second, demographic changes concern the ratio between rural and urban areas, to the benefit of the latter (Figure 10.3). Increased urbanization will have consequences on land use, transport and agricultural production. Finally, the decline in population growth will be accompanied by aging and an expected change in public health policies.

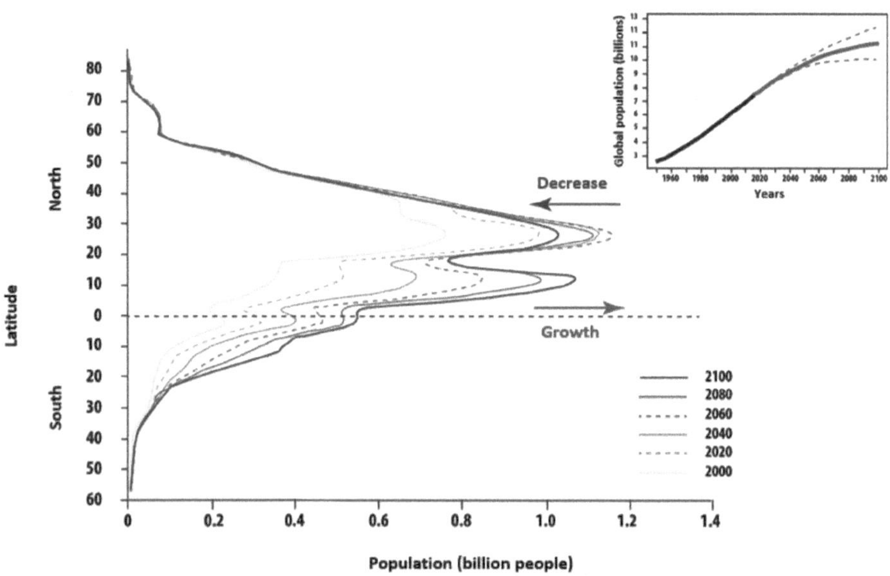

Figure 10.2. *Projections of changes in world population growth according to latitude (taken from McDonald et al. [MCD 13], based on the 2010 United Nations data). There is a decrease in population in the temperate zones of the Northern hemisphere and an increase in population in the tropical zones across the globe. The top right graph gives estimates of population growth by 2100 [UNI 15]. For a color version of the figure, see www.iste.co.uk/morand/biodiversity.zip*

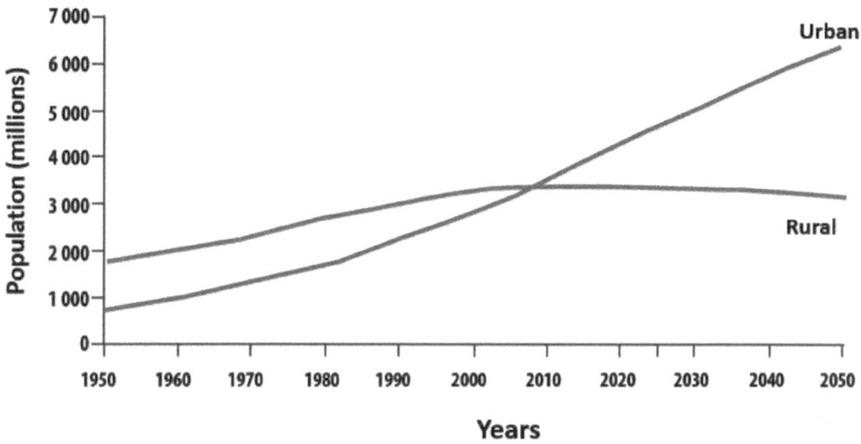

Figure 10.3. *Projection of rural and urban populations (from [UNI 14]). Global growth will affect urban populations. For a color version of the figure, see www.iste.co.uk/morand/biodiversity.zip*

The consequences on biodiversity of population growth in urban areas have been analyzed by McDonald *et al.* [MCD 13], projecting population growth data for the world population on the 34 biodiversity "hotspots" [MYE 00] (Figure 10.3). Over half of the biodiversity hotspots will be significantly affected by an increase in urbanization by 2050, with consequences on biodiversity in terms of urban infrastructure and road communication networks ([IBI 16], Chapter 5). A direct impact of cities on biodiversity relates to changes in land use associated with urban growth. Urban areas currently occupy about 3% of the land area [MCG 06] and urban growth will be an important factor in land-use change with increased deforestation and loss of agricultural land [MCD 13].

Bloom *et al.* [BLO 08] noted a strong correlation between the level of wealth of a country and the proportion of its population that lives in urban areas. These areas do indeed offer richer market structures and higher productivity than the rural world. Bloom *et al.* [BLO 08] pointed out that rapid urbanization is associated with environmental degradation. The lack of correlation between increased urbanization and economic development [BLO 08, CHE 14] suggests that this increase contributes to greater pressure on the environment, and hence on biodiversity, while not providing the economic resources needed for public health systems.

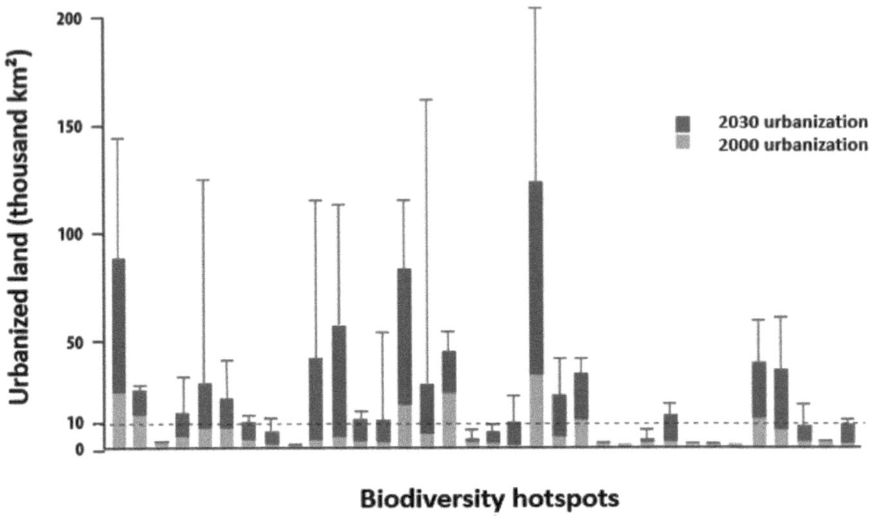

Figure 10.4. *Urban expansion projected in the 34 biodiversity "hotspots" by 2030. Over half of these hotspots will be significantly affected by the increase in urbanization (redesigned based on McDonald et al. [MCD 13]). For a color version of the figure, see www.iste.co.uk/morand/biodiversity.zip*

The environmental impact of urban growth can be estimated through the increase in energy consumption seen in urban buildings. Globally and by 2050, this increase in energy consumption for heating and cooling is estimated to be between 7% and 40% compared to 2010 [GÜN 17]. Increased urbanization will affect renewable and non-renewable sources of energy, as well as carbon sinks and biodiversity [SET 12a].

In addition to the expansion of urbanization, significant changes are expected in agricultural and forest areas [DAN 15, BRE 17]. The results obtained from models based on the extrapolation of historical trends predict a more or less significant increase in agricultural areas according to development scenarios: maintaining current economic globalization and sustainable development (Figure 10.5). Forest areas are particularly affected by economic globalization.

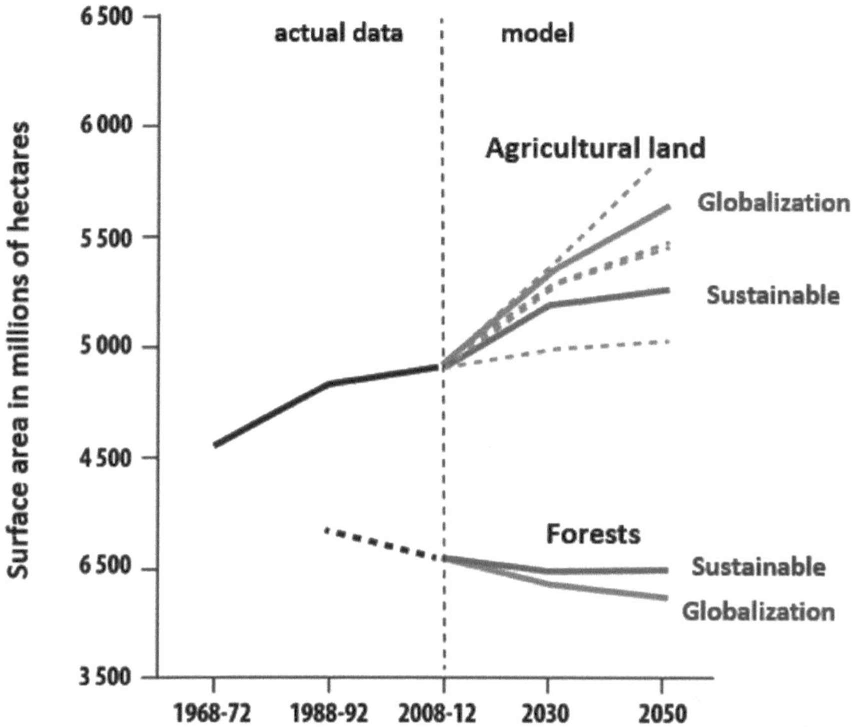

Figure 10.5. *Evolution of agricultural and forest land cover between 1970 and 2050. Data on changes between 2010, 2030 and 2050 are derived from modeling in various scenarios: the scenario of economic globalization is in blue, sustainable development is in green (simulation values are given by the dotted lines) (from [OEC 15, FAO 12] historical data). For a color version of the figure, see www.iste.co.uk/morand/biodiversity.zip*

10.2.2. *Agriculture and livestock*

The increase in world population has been accompanied by an increase in production of cereals and meat (see also Chapter 5 and Figure 5.4 on the evolution of different animal productions over past decades). Here also, the values derived from modeling are highly dependent on economic scenarios. The scenario of economic globalization is accompanied by significant production of animal meat.

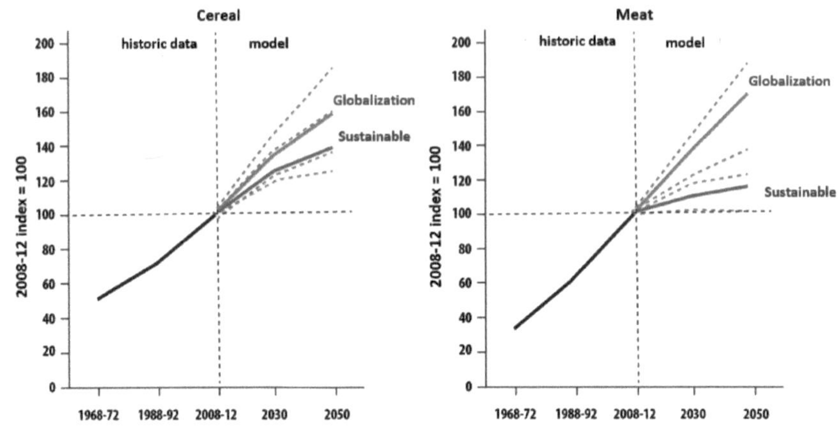

Figure 10.6. *Evolution of cereal and meat production. Historical data are based on physical quantities, while the modeled data are based on production volumes valued at constant prices (2008–2012). The range of the simulation values is given by dotted lines (from [OEC 15]). For a color version of the figure, see www.iste.co.uk/morand/ biodiversity.zip*

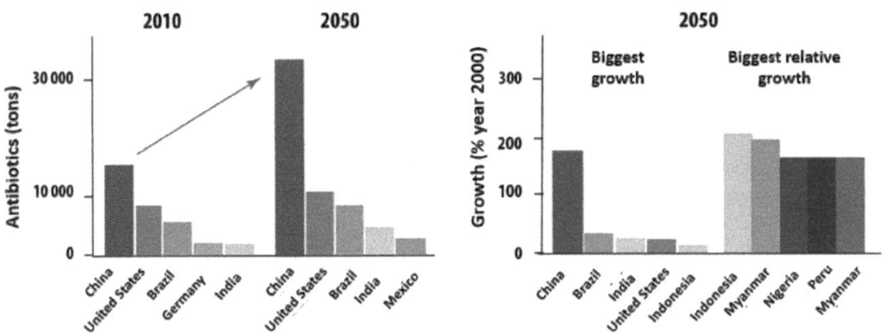

Figure 10.7. *Evolution of antibiotic consumption in livestock between 2010 and 2050 for the biggest antibiotic-consuming countries. Data shown in tons (left) and percentage increase (right) in absolute or relative data, showing the countries that will present the greatest increases in use of antibiotics in animal husbandry (taken from [VAN 14]). For a color version of the figure, see www.iste.co.uk/morand/ biodiversity.zip*

This production of animal proteins requires significant energy resources, in terms of agricultural areas needed for animal feed (see also Figure 5.3), inputs, pesticides, and veterinary health products such as vaccines and antibiotics (Figure 10.7, [VAN 14, PRI 15]). Increased use of antibiotics can thus be extrapolated from existing data. Consequences on the emergence of resistant bacteria at the environment/livestock/human health interfaces therefore need to be considered (see Chapters 5 and 6).

10.2.3. *Climate change*

Economic scenarios serve as a basis for future climate models [IPC 13], taking into account estimates of greenhouse gas emissions from urbanization, transport, livestock and land use (carbon sinks or sources). The projections obtained by synthesizing different models show an increase in average global temperatures, although the amount of change depends on the economic scenario.

The effects of climate change on changes in the distribution of living organisms and on the phenology of organisms can be observed [LOV 05, PEC 17]. Projections of the effects of climate change on infectious diseases also concern vector range changes [DEL 08, ALT 13]. Published models show that the distributions of many infectious diseases will really change, especially for diseases that require the presence of vectors to ensure transmission of infectious agents. These models predict new territories that are at risk due to changes in climatic environmental niches, which will become favorable for the establishment of infectious cycles [GUI 11]. Models generally show a shift in the environmental niches of infectious diseases with displacements from distribution areas towards higher latitudes (as for dengue) or higher altitudes (as for malaria).

Statistical congruences between distributions of pathogens (or their vectors) and changes do not demonstrate a direct impact of climate change on the incidence of infectious diseases [LAF 09, GET 10], bearing in mind that most of the studies are based on modeling scenarios [MAY 15]. The transmission of an infectious agent also depends on local biodiversity conditions, which requires building models that integrate changes in biodiversity with climatic variables [GAL 09].

Climate variability has had a major impact on the history of civilizations [FAG 09]. Climatic phenomena such as El Niño/La Niña are known for their consequences on many infectious diseases. Abnormal events of extreme rainfall favor vector or reservoir-borne diseases such as dengue fever, Japanese encephalitis, malaria or hantavirus hemorrhagic fever [ANY 12].

Inter-annual climate variability, as measured by indices such as ENSO (El Niño South Oscillation) or NAO (North Atlantic Oscillation), is linked to the impacts of several infectious diseases. A temporal and spatial correlation can be seen between the values of these climatic variability indices and the incidence of leptospirosis or dengue in Southeast Asia, or those of hantavirus hemorrhagic fever in Europe [MOR 14]. Teleconnection predicts the incidence and outbreaks of many more infectious diseases using these indices [ANY 12, MOR 14a].

Recent climate models suggest that climate change is changing the intensity and frequency of climate variability [CAI 14]. The El Niño/La Niña events will be more intense in the coming decades. Models predict that the monsoon routine on which African and Asian agriculture depends will see a decrease in volume of average annual rainfall and there will be a greater number of abnormal years characterized by intense droughts or floods. We could assume that the epidemiological environment will be affected by this variability, resulting in an increased risk of epidemics for all water-borne diseases transmitted by vectors or dependent on wild reservoirs, thus affecting humans, domestic animals and wildlife.

Models suggest that climate change will increase the number of extreme weather events and flood risks in many parts of the world [HIR 13]. Combined with other environmental change factors, the risks to water security and biodiversity are particularly significant in the intertropical regions of the world [VÖR 10].

10.2.4. *Biodiversity*

Estimated past and current extinction rates and projections of extinction rates for the 21st Century have been summarized by the CBD (Convention on Biological Diversity; [LEA 10]). These rates are estimated for the distant past (fossil data), for recent times (IUCN Red List data) and for the future following different global environmental scenarios: for birds [JET 17], for the period between 2000 and 2050 based on scenarios on change in land use), vascular plants ([VAN 06], for the period between 1995 and 2050 for four scenarios of global change), and various plant and animal taxa ([THO 04] for the period 2000 to 2050 in relation to scenarios on climate change, Malcolm *et al.* [MAL 06] for the period 2000 to 2100) (Figure 10.8). The models give projected extinction rates with high uncertainties, all of which are higher than the current extinction rates, which are themselves well above the extinction rates of the fossil record.

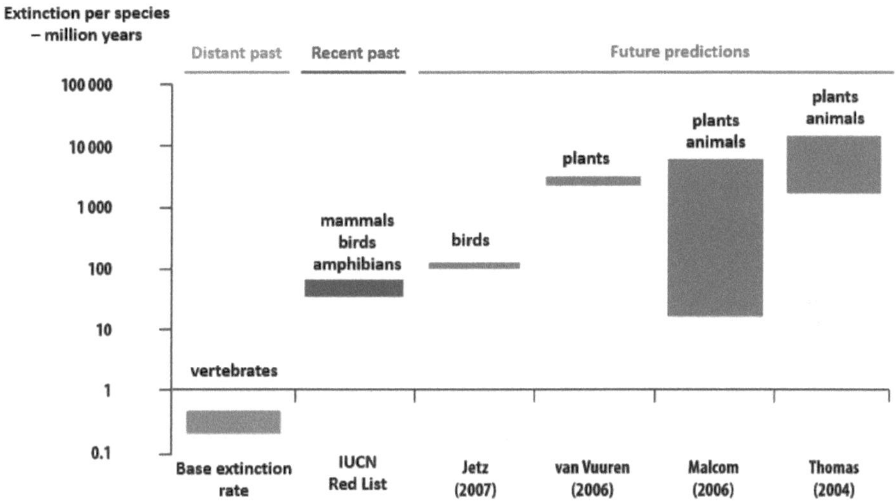

Figure 10.8. *Estimated extinction rates for the past (fossil record), recent times (IUCN Red List data) and for the 21st Century. These rates are estimated as extinctions per million species-years. Different global scenarios are presented for the future: for birds ([JET 17], for the period 2000 to 2050), vascular plants ([VAN 06], for the period 1995 to 2050), and various plant and animal taxa ([THO 04], for the period 2000 to 2050, [MAL 06], for the period 2000 to 2100) (taken from [LEA 10]). For a color version of the figure, see www.iste.co.uk/morand/biodiversity.zip*

The Convention on Biological Diversity (CBD 2010) has set itself the 20 Aichi targets for biodiversity by 2020 (Chapter 11). Tittensor *et al.* [TIT 14] evaluated a set of indicators for progress against indicators associated with the Aichi objectives. The authors projected the trends of these indicators to 2020 using a statistical method that incorporates the specific properties of an individual time series for these indicators. The conclusions were that when public policy efforts are faced with a biodiversity crisis, they are not in a position to halt the negative trends estimated for 2020. Biodiversity pressure variables (human appropriation of biological productivity), the status of biodiversity (Living Planet Index) and responses with biodiversity benefits (number of domesticated breeds, Red List of pollinators) show a foreseeable decline in biodiversity with negative consequences on resources and ecosystem services (Figure 10.9).

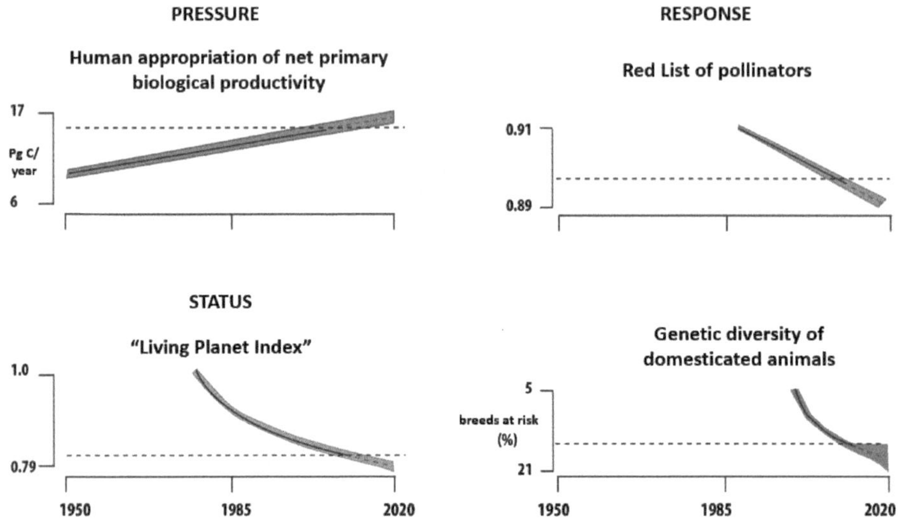

Figure 10.9. *Projections to 2020 of four indicators for Aichi objectives: pressure on biodiversity (human appropriation of net biological productivity), status of biodiversity (Living Planet Index) and responses in terms of benefits linked to biodiversity (terrestrial domestic animal breeds, Red List of pollinators). The model corresponds to the red lines with 95% confidence intervals (colored areas). Significant differences with the 2010 estimates are given by the horizontal dotted line (from [TIT 14]). For a color version of the figure, see www.iste.co.uk/morand/biodiversity.zip*

10.2.5. *Human health*

Each decade in the last century is characterized by an important epidemiological transition with a shift from infectious disease-related human health to non-communicable diseases (Chapter 4) [MAT 06]. This transition was accompanied by a reduction in mortality (infant and adult) and a reduction in fertility, the consequences of which are highlighted in the WHO projections (see section 10.2.1). The Millennium Development Goals aim to speed up the decline of infectious diseases in low-income countries.

Dye [DYE 14] provided projections on the causes of human deaths by 2050. The projections show a significant decrease in deaths from infectious disease, from 16 million deaths in 2010 to 13 million deaths per year by 2050. However, this is still far from total eradication of infectious diseases. The projections predict an explosion in deaths from non-communicable diseases from 31 million in 2010 to 83 million deaths per year by 2050. Such projections question the capacity of public health systems to cope with both the impacts of communicable diseases and the increase in non-communicable diseases.

The effects of environmental changes on human health have been analyzed in a conceptual framework proposed by Myers *et al.* [MYE 13]. On the basis that these changes can provide both benefits and disadvantages for human health, such as hydraulic development (dams, irrigated agricultural perimeters), the analysis showed that the levels of infrastructure (engineering) and access to the market economy prevail in the evolutionary trajectories of the population's health level relative to environmental degradation (Figure 10.10).

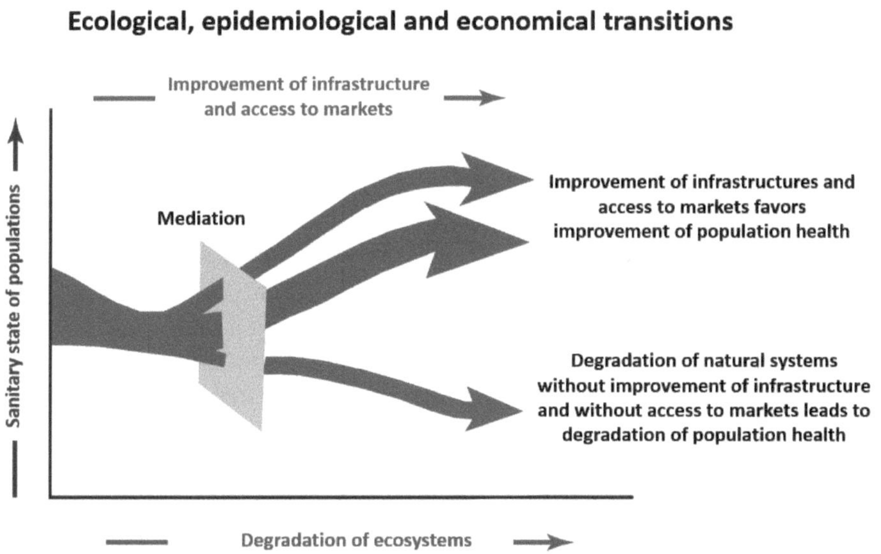

Figure 10.10. *Conceptual diagram of the link between population health and ecological transition, adapted from Myers et al. [MYE 13]. People are moving from a primary dependency on natural systems for health-related ecosystem services to a state where these populations depend on engineering infrastructure for these services and market access. The trajectory of transition depends on many factors that can alter the vulnerabilities and health levels of populations such as economic equity, quality of governance and environmental characteristics (mediation space) (modified from [MYE 13]). For a color version of the figure, see www.iste.co.uk/ morand/biodiversity.zip*

The point of the conceptual framework from Myers *et al.* [MYE 13], in addition to identifying research gaps, is to integrate the dimension of human health into decision-making and governance in the areas of land-use planning and environmental conservation.

10.2.6. *Animal health*

The level of animal health affects the level of health and well-being of human populations, as well as the economic level [GRA 12]. Estimating the burden of infectious diseases that will affect livestock depends on the capacities of animal health systems, which vary widely depending on the economic development of countries. Despite knowledge biases, a study by Perry *et al.* [PER 13] showed that animal infectious diseases decline in rich countries following the trends of human infectious diseases. On the contrary, poor countries continue to suffer the impact of endemic infectious diseases and major epidemics, such as recurring epidemics of Rift Valley fever disease. Infectious diseases that affect animals are more intensively studied in developed countries, with the exception of diseases with pandemic potential for which the agents come from developing or emerging countries (such as H5N1 avian influenza). Projections of this are difficult to achieve, but Perry *et al.* [PER 13] identified three future trajectories: a growing concern in developed countries for disease surveillance and control; intensive and market-oriented livestock production systems in many developing countries, where health risks will be significant; traditional production systems in poor countries for which the diseases will remain neglected.

10.3. Worst-case scenarios

10.3.1. Thresholds and tipping points, planetary limits

Rockström *et al.* [ROC 09] defined thresholds as nonlinear transitions in the functioning of human/environment coupled systems. The thresholds are therefore intrinsic characteristics of systems (Figure 10.11). Not all processes are associated with threshold effects, such as regional or continental land-use change (for a critical threshold on a local scale, see Kéfi *et al.* [KÉF 14]). Nonlinear changes from a desirable state to an undesirable state can affect key ecological functions, leading to major functional collapses. Thresholds and tipping points are therefore states that can be determined by mathematically studying human/environment coupled systems.

Planetary limits are human values defined to maintain a control variable at a "safe" distance from a hazardous level, a risk zone that encompasses thresholds and tipping points ([ROC 09]; Figure 10.11). Determining a safe distance involves normative judgment, which companies choose when faced with risk and uncertainty. Planetary limits can thus be affected by global processes (climate change) or can be aggregated from regional to global (Figure 10.11).

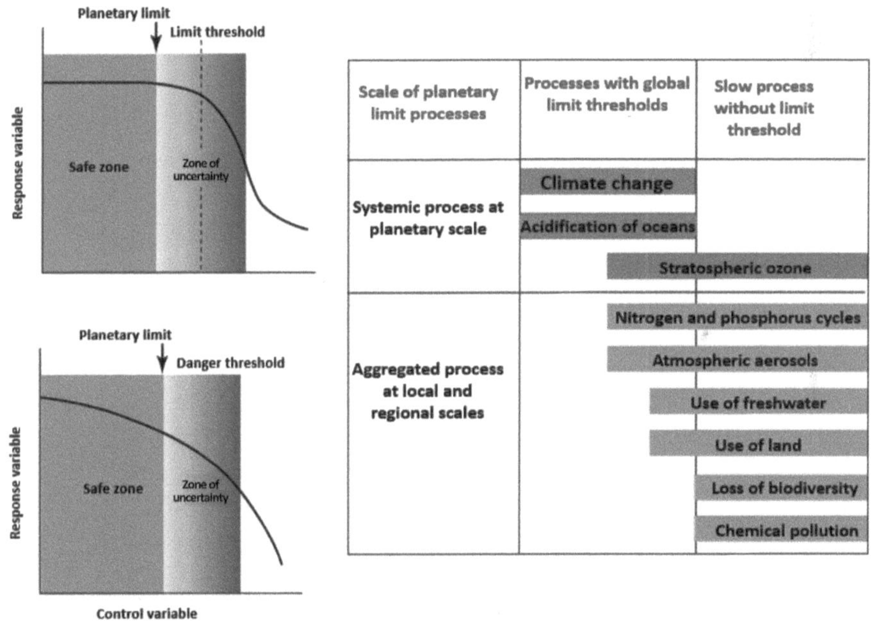

Figure 10.11. *Description of planetary thresholds and limits according to Rockström et al. [ROC 09]. The planetary limit is designed to be a boundary beyond which a critical threshold at the local, regional or global scale affects a process in the terrestrial system (climate change, land-use change), as given in the table on the right. Insufficient knowledge of threshold dynamics generates an uncertainty zone and positions the planetary limit. Exceeding the planetary limit causes threshold effects, which can lead to tipping points. For a color version of the figure, see www.iste.co.uk/morand/biodiversity.zip*

An application of the concept of threshold and planetary limits was given by Pelletier and Tydmers [PEL 10] for animal production systems and their effects on environmental changes. By comparing contributions from the global livestock sector in 2000 with estimated contributions from the sector in 2050, Pelletier and Tydmers [PEL 10] highlighted three important environmental consequences of livestock on: climate change, nitrogen mobilization and appropriation of plant biomass on a global scale. Environmental sustainability requires animal production systems to meet critical thresholds and limitations for these three effects. The proposed scenarios illustrate various expected impacts that are associated with companies' choice of diets. Pelletier and Tydmers [PEL 10] highlighted the magnitude of estimated impacts in relation to the limiting condition of sustainability, which requires prioritizing the livestock sector within global environmental governance (see also [TIL 11]). Let us note that this study does not take into account the indirect effects of livestock on human health.

10.3.2. *Collapse*

The reason for the collapse of societies or civilizations has turned historians [MCN 76, TAI 88, CLI 14] into natural scientists and environmental scientists [DIA 05, SMI 08, TUR 14]. All these studies highlight the cumulative effects of health and environmental crises on societies, which are often characterized by high levels of organization and complexity but for which resilience is too low to overcome the crises. The significance of health crises in the risk of collapse has always been emphasized [GAR 01]. The Ebola epidemic in West Africa in 2015 illustrated the difficulty of containing an epidemic in countries with health systems that have been weakened by different economic crises and in some cases, civil war.

Motesharrei *et al.* [MOT 14] constructed a model of human population dynamics by incorporating accumulated wealth and economic inequality into a predator–prey model, in other words the use of natural resources by humans. The model is simple with four equations that describe the demographics of "elites", "common people", nature and health. The model incorporates an economic stratification and an ecological constraint. Collapse can be avoided if the degree of depletion of nature is limited and inequality is reduced. The simplicity of the model has obviously been criticized, including by the authors, but the results of simulations support the importance of defining planetary limits to avoid tipping points and collapse [COS 07]. Interestingly, the model emphasizes the importance of economic inequality in the risk of collapse, which is negatively correlated with the subjective value of well-being (Chapter 8, Figure 8.1).

10.4. Global risks and "preparedness" for the worst

Worst-case scenarios have direct implications for state governance and particularly for national security.

A report commissioned by the Institute of Medicine of the US Academy of Sciences in 1991 to the Committee on Emerging Microbiological Threats to Health highlights the unrealistic likelihood of a victory over the multitude of existing microbial diseases (and those yet to emerge), and the impossibility of predicting the temporality and spatiality of new infectious diseases that are certain to emerge [LED 92]. While recognizing the coevolving nature of the global ecology of emerging diseases, the committee highlights the difficulty of establishing projections.

In the early 2000s, the CIA published a report on the risks of infectious diseases and bioterrorism for national security, following the 2001 attack on New York and letters laced with anthrax, with the adoption of a law on biological threats by the American Congress. A Pentagon report alerted the Federal Administration of the need to seriously consider climate change and the resurgence of infectious diseases, which were presented as an imminent threat to national security [COO 06a].

The United States believes in preparing for emerging infectious diseases, including bioterrorism, under situations of maximum uncertainty. These are "preparedness" and "worst-case scenarios", which have been developed by successive US administrations [ZYL 13, ZYL 16]. They involve considering that emerging diseases, and terrorism, are driven by similar processes of coevolving races. The war on microbes, like the war on terrorism, would be deemed to be permanent. This amounts to a perpetual war that is declared against whoever integrates the evolutionary processes of living organisms [COO 06a].

More recently, in an unclassified report to the US Senate (Daniel R. Coats, May 11, 2017, Statement for the US Intelligence Community, Senate Select Committee on Intelligence), the Director of US National Intelligence summarized the global threats to US security. Among these threats, the report mentioned wildlife trafficking and illegal fishing, climate change, loss of biodiversity, increased antibiotic resistance and pandemic-risk infectious diseases. According to the report, the links between poaching, the illegal wildlife trade, instability, corruption, crime and challenges to the rule of law are clear. The potential collapse of global fisheries and especially the increase in illegal fishing threaten food and economic security. Illegal fisheries benefit transnational crime and trafficking of human beings and undermine efforts to implement sustainable fisheries policies. Countries with high populations in their coastal areas are particularly vulnerable to climate change and extreme tropical weather events. The degradation of air quality could provoke protests against authorities, such as those seen in recent years in China, India and Iran. Tensions over shared water resources are sources of potential conflict in some parts of the world. Biodiversity will continue to decline as a result of habitat loss, overexploitation, pollution, disruption of ecosystems that support life, including human life. The risks of emergence or re-emergence of highly pathogenic microbes (avian influenza, MERS-CoV) and the report citing World Bank estimates for the cost of a global influenza pandemic could cost the equivalent of 4.8% of global GDP ($3 billion). Finally, it suggests increased resistance to antibiotics, which is likely to exceed the development of new drugs with drug-resistant forms threatening progress in the control of diseases such as tuberculosis or malaria. A recent study showed a link between the level of governance and corruption and the level of antibiotic resistance [COL 15].

10.5. Towards integrated scenarios

The IPBES [IPB 16] highlights the need to build integrated models for societies/biodiversity/ecosystem services by addressing links between ecosystem services through studying the interactions between land-use change and biodiversity [NAG 13].

It is not possible to present all the approaches and studies in this field (see [IPB 16] and Chapter 9), but we can focus on those that explicitly address the links between biodiversity and health.

The social ecology approach presents this application by explicitly linking ecological/biological metabolism with social metabolism [FIS 16]. Social systems are seen as hybrid systems between cultures (communication exchanges) and environments (metabolic exchanges). Human societies are characterized by stocks and flows that involve: population and its demography; biophysical stocks (infrastructure, livestock) and trade/production; land and its biological productivity.

An important parameter is HANPP (Human Appropriated Net Primary Productivity). This has doubled in the 20th Century [KRA 13] and the scenarios presented below suggest that this appropriation will continue to increase considerably in the coming decades. Human appropriation of primary production (HANPP) is an appropriate indicator for research into the impact of human intervention on biodiversity [HAB 07, PLU 16]. This aggregate indicator measures the impact of land use on energy availability (net primary production) in ecosystems and links human activities, such as agriculture or urbanization, to ecosystem processes. Its usefulness lies in directly linking human appropriation of environmental metabolism with ecological theories on biodiversity, like the species-energy hypothesis [VIT 86, WRI 83, WRI 90].

A conceptual framework has been put forward with a dynamic interactive network that links "drivers" (food consumption, energy), pressures (land use, HANPP), states (biodiversity change and extinction of species), impacts (reduction of the quality of ecosystem services) and responses (governance, land planning, conservation) (Figure 10.12). This conceptual framework makes it possible to address the link between biodiversity and health. The species-energy hypothesis is valid at the country level with a positive correlation between HANPP and declining biodiversity. The hypothesis of a reduction in ecosystem services for the regulation of infectious diseases seems to be confirmed with an increase in zoonotic disease epidemics and an increase in HANPP (Figure 10.12).

Social ecology has similarities with the co-viability approach, with the latter aiming to combine social and environmental dynamics. However, a shift to a territory scale is needed. Methods for producing scenarios coupled with landscape changes – changes in biodiversity [EWE 13] – or changes in the provision of ecosystem services are developing [NAG 13], although regulation of infectious and non-infectious diseases are still underdeveloped.

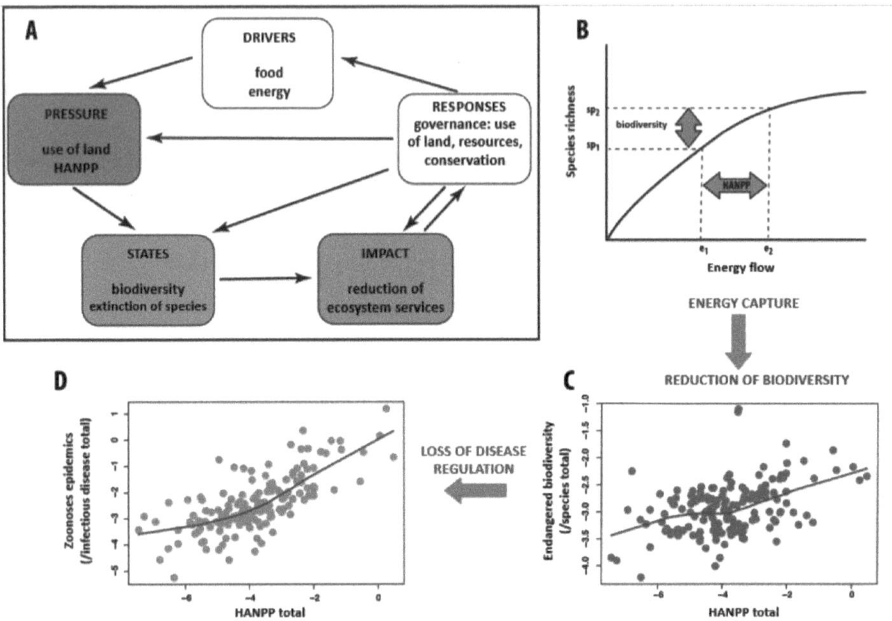

Figure 10.12. *Conceptual framework of dynamic interactive network that links "drivers" (food consumption, energy), pressures (land use, HANPP), states (biodiversity change and extinction of species), impacts (reduction of quality of ecosystems) and responses (governance, land planning, conservation) (modified from [PLU 16]). This framework is based on the species-energy hypothesis [VIT 86, WRI 83, WRI 90], where a decrease in biological productivity, notably through human appropriation (HANPP), translates to loss of biodiversity (Δ biodiversity). This conceptual framework finds empirical support in the correlations which we observe between HANPP ([IMH 06] data, http://sedac.ciesin.columbia.edu/ data/collection/ hanpp) and the number of endangered species (IUCN Red List data), and between HANPP and the number of zoonotic disease epidemics (data from GIDEON, [MOR 14a]). For a color version of the figure, see www.iste.co.uk/morand/ biodiversity.zip*

A conceptual framework (Figure 10.13) that combines retrospective modeling and future scenarios of land-use changes and their health consequences can thus be proposed. As economic aspects are one of the elements of mediation [MYE 13; Figure 10.11), local and national governance must be considered as the essential driver for land planning (agriculture, conservation, urbanization) and public health (including veterinary health). Retrospective modeling based on phenomenological/ statistical models allows us to analyze and highlight determinants and interactions between factors of change and consequences on public health.

Prospective scenarios are based on models that incorporate processes (such as multi-agent models, epidemiological models, etc.) and, depending on external conditions (climate change, global economy, demography), produce potential scenarios of changes in land use through mediation of local/national governance. These scenarios, which can be "worst-case" or "best-case" (i.e. desirable), in turn enter mediation by affecting various elements of governance that will lead to new scenarios. The process is therefore iterative and requires close collaboration with different actors in the socio-ecosystem.

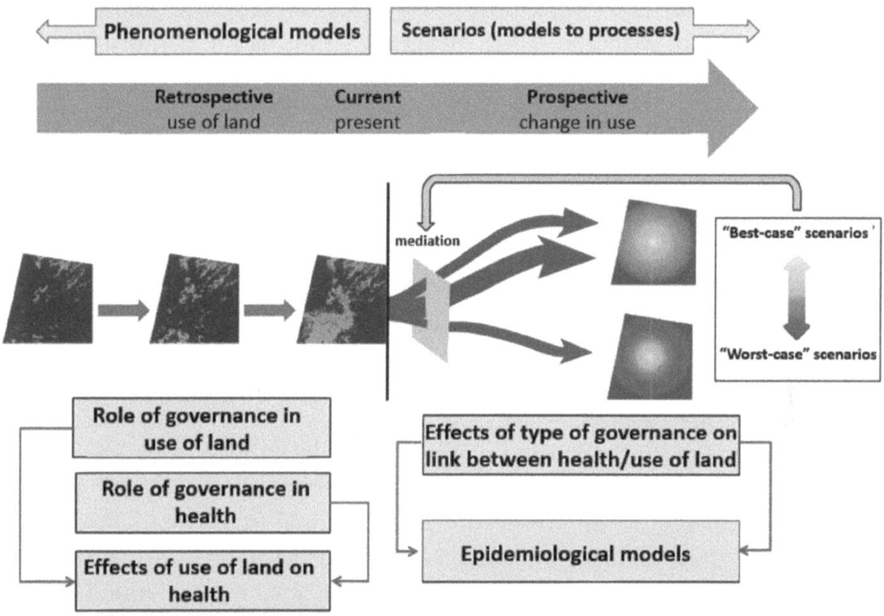

Figure 10.13. *Conceptual framework to produce retrospective modeling and predictive scenarios for the links between land use (agriculture, conservation, urbanization) and public health (including veterinary and phytosanitary). Retrospective modeling that is based on phenomenological/statistical models allows us to analyze the determinants and interactions of factors of change and their consequences on public health. Prospective scenarios are based on models that incorporate processes (such as multi-agent models, epidemiological models, etc.), which integrate external conditions (climate change, global/local economy, socio-demography) and produce potential scenarios for land-use change in the mediation of local/national governance. These "worst-case" or "best-case" (i.e. desirable) scenarios are incorporated into mediation to produce new scenarios. The process is iterative and requires close collaboration with the different elements in the socio-ecosystem. For a color version of the figure, see www.iste.co.uk/morand/biodiversity.zip*

Various public health elements can be incorporated, such as well-being, mental health, infectious diseases and autoimmune diseases (allergies, diabetes) [LID 16].

10.6. Observations and observatories

International agencies, as mentioned in previous chapters, provide many socio-economic, cultural, environmental and health data (WHO, FAO, OIE, World Bank, OECD, UNICEF, IUCN, WWF, BirdLife, etc.). Most of these data are aggregated across countries. Biodiversity data are georeferenced more accurately through the GBIF program (http://www.gbif.org/) and its online databases. Recently, the PREDICTS project (Projected Responses of Ecological Diversity in Changing Terrestrial Systems, www.predicts.org.uk) compiled a large database of comparable biodiversity samples from multiple sites that differ according to the nature or intensity of human impact linked to land use [HUD 17]. The 2017 version of the database contains over 3.2 million samples from 26000 sites, representing over 47000 species. This database analyzes patterns of biodiversity on local, regional and global scales.

The need to collect long-term data in a standardized manner in order to develop scenarios for eco-evolving dynamics of zoonoses and their reservoirs and vectors is the core of the NEON initiative ([SPR 16]; Figure 10.14). In the light of the scarcity of long-term chronological data on infection rates in vectors and reservoirs, the National Ecological Observatory Network (NEON) collects measurements and samples on vectors and zoonotic diseases on a continental scale (the territory of the United States). Springer *et al.* [SPR 16] described sampling models and sampling priorities, field and analytical methods, and management of archived data and samples that will be made available for research. The knowledge generated by this sampling will help to better understand and predict changes in dynamics of zoonotic diseases in an interdisciplinary and collaborative manner.

A key source of biodiversity information is remote sensing [SEC 14, WAL 17], which supplements other sources of biodiversity data [PRO 16]. The integration of spatial data into geographic information systems allows us to develop statistical models for changes in biodiversity [MOU 15] and in vectors and reservoirs of infectious agents [PET 14].

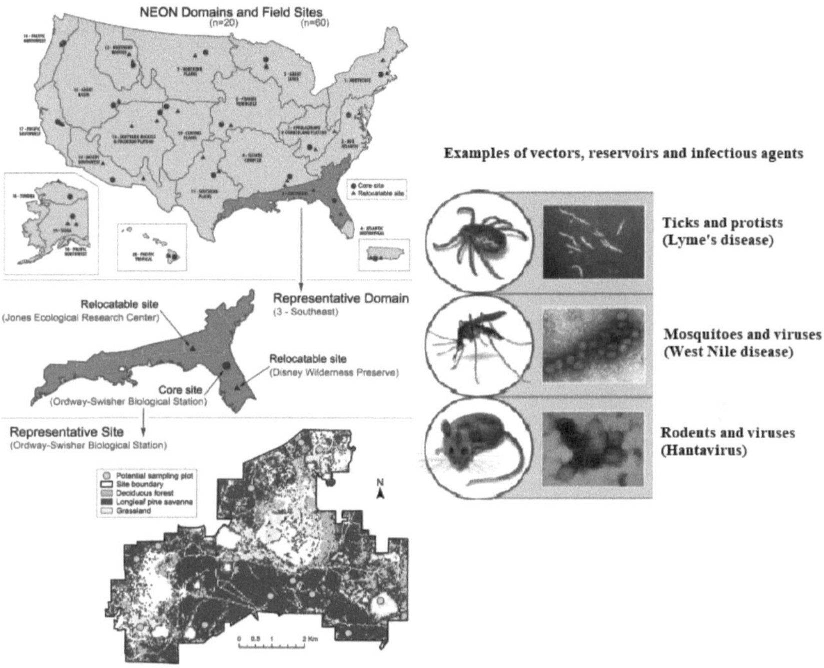

Figure 10.14. *NEON and zoonotic diseases [SPR 16], based on the hierarchical spatial structure of NEON (National Ecological Observatory Network) [SPR 16]. On the left, top to bottom: boundaries of the 20 eco-climatic domains and the locations of associated sites; example distribution of a central site and two sites in a representative domain with multiple sampling plots distributed among the vegetation types. On the right: a network of the collections and sampling of vectors and reservoirs and infectious agents (images from the CDC Public Health Image Library in the United States, from [SPR 16]). For a color version of the figure, see www.iste.co.uk/morand/biodiversity.zip*

Citizen science [BOA 16] and participatory epidemiology [ALL 17] are constantly developing. In addition to making valuable data available to the scientific community, citizen science enables the science/society dialogue to be developed within the science/policy dialogue and ultimately reinforces the rigor and relevance of the scientific approach [BAL 13]. In addition to methods such as the "Health Impact Assessment" (see [LAJ 15a]), participatory methods can better analyze perceptions of health and environmental risks in terms of local knowledge and local governance [STI 09, FIS 16].

10.7. Experts and representation of knowledge

Scenario co-construction is the preferred tool for decision support [NAS 17]. This co-construction is based on participatory methods and brings into play the effects of individual and collective decisions on a future that human and non-human communities want and (must) share. The "health" or "conservation" service can thus be represented at different spatial and temporal scales. Mapping these services makes it possible to represent land use conditions (conservation, water regulation, carbon sequestration) that are directly linked to these services. Innovative participatory methods have been developed for these mappings [NAS 17].

However, some difficulties have been encountered in the representation of disease control services. For example, if the regulation of both communicable and non-communicable diseases is based on a habitat fragmentation index that is associated with forest cover, the service will be difficult to map directly. It will require a recommendation on the level of fragmentation of forest areas. We also emphasize that mapping presents the danger of "freezing" the landscape and losing an adaptive overview of landscapes.

It is crucial to communicate scenarios and models to policy-makers, professionals and local communities [SUT 12]. Thus, it involves communicating consensus, uncertainty and controversy. The link between biodiversity and health is an example of communication problems in terms of scientific expertise. There is a good scientific consensus on the erosion of biodiversity, epidemiological transition and emergence of non-communicable diseases, and the emergence of new epidemiological risks such as emerging infectious diseases or antimicrobial resistance. On the contrary, scientific uncertainties are numerous, as they are the effects of land-use changes on epidemiology. These uncertainties must be resolved through new studies. Finally, there are controversies such as the dilution effect, which occurs due to epistemological aspects (the nature of theory and associated hypotheses) and ideological and political aspects (access to funding for scientific projects).

10.8. Conclusion: scenarios for research and governance

As highlighted by Naem *et al.* [NAE 16], biodiversity and human well-being are two essential components and are linked to each other to achieve sustainable development goals. Responses to health crises have implications for governance, from a local to a global level. This chapter has illustrated the diversity of scenarios for expected changes in demographics, urbanization, agriculture, land use and their implications for biodiversity and health linkages.

The production of scenarios requires observatories and observations and their integration into models that take into account the effects of scale changes that affect socio-economic and ecological processes.

Scenarios become meaningful when they are co-constructed with different elements from socio-ecological systems (users, managers, politicians). This requires a dialogue between science, society and politics, where perceptions and representations of scientific expertise are of central importance.

The construction of predictive ecology and epidemiology has important implications for governance, which takes scenarios of scientific expertise into account (Chapter 11). It also requires questioning oneself on the role of ethics in scientific practices (Chapter 12).

11

Governance of Biodiversity and Health

11.1. Introduction

The 1968 UNESCO Biosphere Conference (in partnership with the FAO, WHO and the UN) [UNE 68] was the first intergovernmental conference to consider both human and health ecology from a scientific point of view [BOU 93]. Thus, it has a leading role in addressing health issues related to ecological diversity on a global scale. It insists on the need for integrated, planned, multidisciplinary policy action for the use and conservation of natural resources within the framework of international cooperation. It notes that:

> "Political decisions should take into account ecological as well as economic considerations".

This conference was followed by many international conferences and initiatives (see Figure 1.2 in Chapter 1 and Lajaunie *et al.* [LAJ 15]), which began with the United Nations Conference on the Human Environment in Stockholm in 1972, which considered the environment in its economic, political and social aspects. Subsequent to this conference, the United Nations Environment Program (UNEP) was established to coordinate United Nations activities in the field of environment and to assist countries in the implementation of environmental policies.

Among the missions of the organization, the 1946 WHO Constitution (Article 2) mentioned the improvement of environmental health. This mission was reaffirmed at the end of the Stockholm Conference through several resolutions by the World Health Assembly (decision-making body of WHO).

It should be noted that the Stockholm Declaration, in its preamble, affirms that to protect and improve the environment for present and future generations:

> "... responsibility by citizens and communities and by enterprises and institutions at every level, all sharing equitably in common efforts. Individuals in all walks of life as well as organizations in many fields, by their values and the sum of their actions, will shape the world environment of the future. Local and national governments will bear the greatest burden for large-scale environmental policy and action within their jurisdictions. International cooperation is also needed...".

These different levels of decision-making, from global to local level, and sometimes both combined, are what we will consider in order to study the governance of biodiversity and health.

11.2. International governance of biodiversity and health

There is no international body that explicitly includes both biodiversity and health in its mandate, but, in this section, we will show how biodiversity and health issues have been indirectly taken into account by international organizations that are concerned with environment and health.

For example, the 1992 United Nations Conference on Environment and Development (UNCED) in Rio provided an opportunity to adopt the Convention on Biological Diversity (CBD, see Box 11.1). It was an international environmental treaty, which has today been ratified by 196 parties with the notable exception of the United States, and is aimed at conserving biodiversity and the sustainability of its components. At the Rio summit, the Director of the WHO also called for a paradigm shift for (human) health and the adoption of a holistic vision to include a physical, social, behavioral and ecological context[1]. Building on the outcomes of the conference, the WHO proposed a global strategy for health and environment (1993), which provided a unified framework for action and the basis for

1 WHO Director General, "A paradigm for health: a framework for new public health action", EB89/11, December 9, 1991.

building the organization's programs on a global, regional and country-wide level.

Seven international conventions focus on biodiversity issues and aim to implement actions at national, regional and international level to achieve common objectives for conservation and sustainability:

– the Convention on Biological Diversity (date of adoption, 1992) https://www.cbd.int/;

– the Convention on the Conservation of Migratory Species (1979) http://www.cms.int/en/;

– the Convention on International Trade in Endangered Species of Wild Fauna and Flora (1973) https://cites.org/eng;

– the International Treaty on Plant Genetic Resources for Food and Agriculture (2001) http://www.fao.org/plant-treaty/overview/en/;

– the Ramsar Convention on Wetlands (1971) http://www.ramsar.org;

– the Convention concerning the Protection of World Cultural and Natural Heritage (1972) http://whc.unesco.org/en/conventiontext/;

– the International Plant Protection Convention (IPPC) https://www.ippc.int/en/;

– Liaison Group on Biodiversity Conventions.

In 2004, the 7th Conference of the Parties to the Convention on Biological Diversity[2] decided to establish a liaison group to connect biodiversity-related conventions with a view to enhancing coherence and cooperation in their respective implementation. This liaison group promoted synergy between the different conventions.

Box 11.1. *Biodiversity-related conventions*

The CBD refers to human health in relation to adverse impacts from environmental change on the conservation and use of biological diversity. However, in 2000, while it was clear that the ecosystems on which human life depends were threatened, the causes and extent of the damage remained to be investigated. The Secretary General of the United Nations therefore

2 CBD, COP7, Decision VII/26, Cooperation with other conventions and international organizations and initiatives.

launched the Millennium Ecosystem Assessment (MA) with the objective of assessing the consequences of ecosystem changes on human well-being and establishing a scientific basis for implementing the actions needed to improve conservation and sustainable use of these ecosystems. The MA synthesis report for health was published in 2005 and called on the health sector to ensure that benefits to human health and well-being from the natural environment, and hence biodiversity, are preserved for future generations [WHO 05].

Simultaneously, regardless of biodiversity issues, the United Nations took note of the globalization of trade (goods, displacement, human and animal interactions) and its effects on health and particularly on the transmission of infectious diseases, and recognized the importance of international and regional cooperation in the control of infectious diseases. It called on Member States to give priority to the revision of the International Health Regulations[3]. In 2005, the outbreak of severe acute respiratory syndrome (SARS) led to the adoption of new International Health Regulations. It was an international legal tool designed to help protect all States from any extraordinary event involving a risk of international spread or requiring coordinated international action [WHO 07]. It entered into force on June 15, 2007. It is currently legally binding for 196 Party States (including all WHO Member States).

The purpose of the new Regulation was to "*prevent, protect against, control and provide a public health response to the international spread of disease in ways that are commensurate with and restricted to public health risks, and which avoid unnecessary interference with international traffic and trade*".

It should be noted that the scope of this new regulation was extended to include any extraordinary event that constitutes a public health emergency of international concern[4], including environmental risks[5].

3 United Nations General Assembly, Resolution 58/3 Intensification of Capacity Building in the Public Health Sector Across the World, 43rd plenary meeting, October 27, 2003.

4 Article 1 defines an event as "a manifestation of disease or an occurrence that creates a potential for disease".

5 WHO, International Health Regulations (2005), Brief introduction to application in national legislation, January 2009, WHO/HSE/IHR/2009.2, p. 2.

In order to cooperate on the effective implementation of the International Health Regulations and improve transparency among countries, the WHO, FAO and OIE (World Organization for Animal Health) agreed in 2006 to coordinate and share information on outbreaks of animal and human diseases and their epidemiological analyses by establishing the Global Early Warning and Response System (GLEWS). In 2010, recognizing that cooperation among countries remained limited in terms of control systems, the three organizations decided to adopt a long-term international collaboration and share responsibility to coordinate management of health risks at the human/animal/ecosystem interface on a global scale. The purpose of the tripartite strategy was explicitly to achieve the objectives of the One Health initiative, to attain "*a world capable of preventing, detecting, containing, eliminating, and responding to animal and public health risks attributable to zoonoses and animal diseases with an impact on food security through multi-sectoral cooperation and strong partnerships*" [6].

A decisive step in integrating the One Health approach into international governance and the legal sphere came from the 2010 Hanoi Conference, which was co-organized by the European Union and the United States in partnership with the FAO, WHO, OIE, Asian Development Bank, World Bank and UNICEF. At the end of the conference, the Hanoi Declaration constituted a commitment by participants to continue in their efforts to improve pandemic risk prevention plans to jointly strengthen animal and human health systems. They recognized the need to align national strategies with regional strategies to address the challenges of implementing the One Health approach at the animal–human–environment interface.

Nevertheless, there had been no clarification regarding consideration of the environment in national strategies. Among the international organizations involved in the fight against infectious diseases, none is specialized in environmental or biodiversity issues.

Consideration of the environmental aspect of infectious diseases emerged with the issue of wildlife health at the 10th Conference of the Parties to the

6 FAO-OIE-WHO Collaboration, A tripartite concept note: Sharing responsibilities and coordinating global activities to address health risks at the animal–human–ecosystems interfaces, 2010.

Convention on Migratory Species (CMS, see Box 11.1), which recognized that the "One Health" initiative was gaining ground as a multidisciplinary means for treating infectious diseases[7] and was specified in the terms of reference of the scientific working group for ecosystem health and wildlife[8]. Integration of health issues into key international agreements related to biodiversity was accelerated through the development of agreements and initiatives between international organizations and between secretariats of the various biodiversity-related conventions (Lajaunie, Mazzega [LAJ 16a] and Box 11.1) and the Rio conventions (CBD, UN Framework Convention on Climate Change, UN Convention to Combat Desertification, see Lajaunie, Mazzega [LAJ 16]).

In 2012, the lack of appropriate policy responses to the deterioration of biodiversity led to the creation of an intergovernmental platform to support policy-making in the area of biodiversity and ecosystem services (IPBES) through tools, methods and scientific knowledge.

While the IPBES's broad thinking framework did take human well-being into account, it did not explicitly refer to the health term that was mentioned during the first meetings of the platform.

It should be noted, however, that the World Organization for Animal Health (OIE) and the Secretariat of the CBD signed a cooperation agreement in 2013 to promote the One Health approach, while in 2015, the CITES and OIE collaboration decided to *conserve biodiversity by ensuring the effectiveness of health surveillance and control measures*, which are required to protect animal and human health worldwide. In addition, the WHO and the CBD Secretariat presented a joint report in 2015 that was entitled "Connecting Global Priorities: Biodiversity and Human Health, Summary of the State of Knowledge Review" [WHO 15a].

---

7 In UNEP/CMS Resolution 10.22, Wildlife disease and migratory species, Bergen, November 20–25, 2011, section 1.

8 Which aims to encourage an integrated scientific approach within the framework of the "One Health" Initiative. From CITES, 26th Meeting of the Animals Committee. Relationship between wildlife trade and wildlife diseases, AC26 Doc. 23 (Rev. 1), Geneva (Switzerland), March 15–20, 2012 and Dublin (Ireland), March 22–24, 2012, Annex section 1 and section 3B.

Finally, the 13th Conference of the Parties to the CBD took note of this joint WHO-CBD publication by highlighting the importance of the One Health approach as an integrated approach for managing ecosystems and the associated human settlements and livestock. This was done in order to minimize unnecessary disturbance of natural systems, thus avoiding or mitigating the potential emergence of new pathogens[9].

11.3. Regional challenges

Determining and addressing regional issues is fundamental to the implementation of internationally defined health and environment (or biodiversity) policies and strategies. Indeed, these challenges constitute the common concerns of Member States from the same region, beyond their diversity, and justify a collective intervention to resolve issues that go beyond the national framework. They highlight issues that are specific to a regional geographical group, whether these involve historical, geographical, cultural, economic or socio-ecological factors.

By taking these regional issues into account, international and regional organizations and a multitude of public and private stakeholders proposed solutions, forming networks of partners to implement the principles defined at international level, and also to try to resolve problems that were common to several States in a given geographical area. These two approaches, one from the international commitments of the States and the other resulting from local problems, constituted the founding elements of regional governance.

At the regional level, the network dynamics, which were formed more or less formally to respond to environmental or health issues, required many actors to create a certain form of flexible organization that could help with the implementation of health and environmental regulations. This emerging network organization, which generated its own rules, was highlighted in a report by the United Nations International Law Commission on the

9 CBD, COP13, Decision XIII/6, Biodiversity and human health, Cancun, Mexico, December 4–17, 2016.

difficulties of fragmenting international law[10] for its networks that emerged from intergovernmental cooperation. For health and environment networks, these could be an asset and offer a new, more flexible form of coordination than international organizations.

To illustrate regional governance and the type of regional policies and initiatives that address regional issues related to health and biodiversity, let us take the example of the Southeast Asia region (as a geographical, biogeographic and political entity), which has a high potential for emergence or re-emergence of infectious diseases.

In Resolution 46.20 of the World Health Assembly[11], the WHO Regional Offices were instructed to use the 1993 Global Health and Environment Strategy to develop regional strategies and action plans. The WHO Regional Offices[12] adapted the global strategy to the specific needs of their region [OZO 94] by consulting the countries concerned. For example, the two WHO offices in the Southeast Asia region adopted health and environment strategies[13] that aimed to work with national and local authorities to create environmental conditions that meet minimum health requirements. These strategies promoted understanding of the interactions between environment, health and development among policymakers and citizens in order to strengthen public policy in these sectors. They planned to improve the capacity for prevention and response to health emergencies, monitoring, managing and evaluating environmental health risks.

10 ILC Analytical Study 2006, ILC Study Group on the Fragmentation of International Law. Fragmentation of International Law: Difficulties Arising from the Diversification and Expansion of International Law; Report of the Study Group of the International Law Commission, Finalized by Martti Koskenniemi. UN Doc A/CN.4/L.682. New York: International Law Commission, 2006, section 481.

11 WHO, Resolution WHA46.20, WHO global strategy for health and environment, May 12, 1993, al. p. 5.

12 There are two separate WHO regional offices covering the countries commonly considered to belong to the geographical area of Southeast Asia: the Southeast Asia Office (SEARO) and the Asia-Pacific Office (WPRO).

13 WHO SEARO, Regional strategic plan for health and environment for the WHO Southeast Asia Region (1993), p. 18 and WHO WPRO, Regional Strategy on Health and Environment, including follow-up of the United Nations Conference on Environment and Development, WPR RC44/13, 1993, p. 16.

Similarly, the Millennium Ecosystem Assessment was carried out as a "multi-scale" assessment and was also conducted at intermediate levels (local, national, regional and river basin levels) to meet the needs of policy makers and improve management of ecosystems. In the Southeast Asia region, these assessments concerned Indonesia, the Philippines and the wetlands downstream of the Mekong River in Vietnam.

The need to share knowledge on good practices and communal efforts in the biodiversity sector between ASEAN Member States[14], in particular in accordance with Article 14c of the CBD[15], led to the creation of a regional institution in 2005: the ASEAN Center for Biodiversity. Hosted by the Government of the Philippines, this institution was intended to be a center of excellence to facilitate regional and international cooperation in the conservation and sustainability of biodiversity. From its inception, the ASEAN Biodiversity Center also developed a clearing house for regional information on biodiversity while contributing, in particular through the organization of workshops and training, to the establishment and development of national information exchange centers. However, these data may be of disparate quality and it is necessary to know how they were gathered, where, by whom and with what degree of accuracy or reliability. Indeed, in addition to these national information exchange centers that were established in accordance with the provisions of the Convention on Biodiversity, many initiatives in Southeast Asia and more broadly in the Asia-Pacific gather data on biodiversity. Among these is the Asia-Pacific Biodiversity Observation Network (AP-BON), a regional network linked to the Global Biodiversity Observing Network (GEO-BON), which collects information on all levels of biodiversity.

In the area of health, in 2005 (the year in which the International Health Regulations were adopted), the Southeast Asia Office of the WHO proposed a plan of action to combat infectious diseases in the region, in light of the public health impact of SARS and avian influenza epidemics as well as to

14 Association of Southeast Asian Nations.

15 "*Promote, on the basis of reciprocity, notification, exchange nf information and consultation on activities under their jurisdiction or control which are likely to significantly affect adversely the biological diversity of other States or areas beyond the limits of national jurisdiction*", CBD, article 14c.

anticipate the emergence of new pathogens[16]. The paper began with the observation that countries in the region had, until then, mainly taken public health measures during crisis situations, so it was high time to set up a surveillance and prevention system. This strategy proposed a key element of developing research activities to improve knowledge about a whole range of infectious diseases and the factors that influence them (such as pathogen evolution, social inequalities, behavior when faced with disease, environmental factors and the impact of environmental change/climate change, changes in land use, disruption of hydrological regimes).

In the regional context of rapid environmental change with reduced biodiversity and increased incidence of pandemic-risk infectious diseases, many networks and initiatives have been created in relation to the environment (agriculture, forestry, land use, biodiversity) and health (infectious diseases, health monitoring, zoonotic diseases). Since its creation, ASEAN has been characterized by its minimally institutional organization, limited use of formal rules and a principle of non-interference in the internal affairs of other Member States [CHA 09]. This has led to the implementation of national regulations, cooperation programs and a system of dialogue within the Member States, and the development of regional programs by non-State actors [LAJ 16c].

The first Asia-Pacific Forum of Ministers and Environmental Authorities meeting in Bangkok in 2015 recalled the importance of examining the links between health and the environment to address urgent environmental health issues and discuss potential policy solutions. The forum strongly emphasized that strategies should focus on promoting the health benefits of healthy ecosystems (whether in terms of food security and nutrition, traditional medicine, mental health and physical and cultural well-being) and the management of ecosystems that prevent their degradation, such as to reduce the risk of infectious diseases[17].

16 WHO SEARO, Combating Emerging Infectious Diseases in the Southeast Asia Region, (SEA-CD-139), February 2005, p. 36.

17 UNEP Regional Office Asia Pacific, Discussion Document for Session 4 Environment and Health Linkages in Asia Pacific, 1st Forum of Ministers and Environment Authorities of Asia Pacific, Bangkok, May 19–20, 2015.

We must also bear in mind that regional initiatives do not imply regional players and an analysis of existing initiatives and networks will provide a clearer picture of a more or less formal articulation of governance (from international to local level).

11.4. Implementation at the national level

The Convention on Biological Diversity is a framework convention that proposes a flexible body of obligations and is legally binding on Party States. These must be implemented through national law and public policy[18]. The decision-making power is thus placed at the national level and each country, as a party to the Convention, interprets the provisions of the CBD according to its national or regional priorities to make them operational. Parties should also develop National Biodiversity Strategies and Action Plans (NBSAPs) and provide information on actions taken to implement the Convention and their effectiveness.

Article 8 of the Convention lays down a list of the measures required to apply the essential elements of conservation *in situ*, which is the preferred means of conservation (protected areas, regulation and management of biological resources both within and outside of protected areas, protection of ecosystems and natural habitats as well as species populations). Article 9 lists the measures necessary for conservation of biodiversity *ex situ*, which should be complementary to *in situ* measures. These include measures to conserve components of biological diversity, preferably in their country of origin; restoration and regeneration of endangered species and reintroduction of these species into their natural habitat under good conditions, as well as regulate and manage the collection of biological resources within natural habitats for *ex situ* conservation purposes to avoid endangered ecosystems and populations of species *in situ*[19].

---

18 On the core characteristics of the CBD (comprehensiveness, complexity, compromise), see McGraw DM, 2002. The CBD – key characteristics and implications for implementation, Review of European Community and International environmental law, 11 (1), pp. 17–28.

19 Glowka L. *et al.* [GLO 94], A Guide to the convention on Biological Diversity. IUCN, Gland and Cambridge, xii + p. 193.

Similarly, the 2005 International Health Regulations have been implemented nationally. States have had to establish or designate a national IHR focal point and national authorities to be responsible for the implementation of IHR measures. National focal points circulate urgent communications linked to the implementation of the IHR to a WHO IHR contact point (an IHR contact point was designated in each of the six regional offices). They also disseminate information to the competent sectors of the party State's administration, including sectors responsible for disease monitoring and reporting, entry points[20], public health services, dispensaries and hospitals, and to gather information from these sectors[21].

States therefore had to respond to IHR obligations and public health monitoring at all administrative levels and with the appropriate means.

The IHR authorizes the WHO to consider informal reports on public health events from public sources other than States (and not only public health information from governments) and obtain verification of these from party States, in accordance with a principle of international solidarity[22]. The WHO must share the information obtained from non-governmental sources with all Member States and intergovernmental organizations if they are required to respond to public health risks of international concern.

These two examples show that international regulation must be effectively implemented at the local level and that this is an obligation of the party States. It remains to be seen how this implementation should be carried out, which depends on the policy choices that underlie the regulation to be defined.

20 Annex 1b in the IHR defines point of entry as "*a passage for international entry or exit of travellers, baggage, cargo, containers, conveyances, goods and postal parcels, as well as agencies and areas providing services to them on entry or exit*".

21 WHO, International Health Regulations, 2nd edition, Geneva, 2006, Annex 1.

22 Heymann D. L., SARS and Emerging Infectious Diseases: A Challenge to Place Global Solidarity above National Sovereignty, Annals of the Academy of Medicine, Singapore 35, no 5 (May 2006), pp. 350–353.

In 2010, the Secretary-General of the United Nations made a statement of failure to implement the CBD and stressed that the main pressures leading to loss of biodiversity were intensifying[23]. The States in the CBD decided to adopt a strategic plan for 2011–2020 with 20 objectives linked to biodiversity, called the Aichi Targets[24]. These objectives express global aspirations and provide a flexible framework to determine national and regional targets. Achieving these objectives requires that "*by 2050, biodiversity is valued, conserved, restored and wisely used, maintaining ecosystem services, sustaining a healthy planet and delivering benefits essential for all people*"[25].

The achievement of these objectives depends first and foremost on common international willpower and implementation at national level through the political will of States as well as constantly balancing divergent interests. For example, protected areas created for conserving biodiversity and ecosystem services raise a number of issues concerning the right of local communities to these protected areas (equity, benefit sharing), the type of resource management[26] (community management, state management or non-governmental organizations) and the preferred mode of management (directive or participatory).

Many initiatives for participatory approaches have been developed at the local level by different types of stakeholders, NGOs, research centers (national or external) and international institutions working in the field of biodiversity, particularly through the conventions linked to biodiversity.

A widely accepted participatory approach saw the light of day in the rural development sector in the late 1970s [CHA 94] with the idea of breaking with policies that had been imposed by decision-making centers and instead

23 Secretariat of the Convention on Biological Diversity, Global Biodiversity Outlook, 3rd edition 2010, Montreal, Foreword by UN Secretary-General Ban Ki-Moon.

24 In accordance with decision UNEP/CBD, IX/9 Process for the revision of the Strategic Plan, October 9, 2008. The plan was adopted by decision UNEP/CDB, X/2, Strategic Plan for Biodiversity 2011–2020 and the Aichi Biodiversity Targets, October 27, 2010.

25 UNEP/CBD, X/2, Strategic Plan for Biodiversity 2011–2020 and the Aichi Biodiversity Targets, October 27, 2010, II, section 11.

26 Borrini-Feyerabend G., Dudley N., Jaeger T., Lassen B., Pathak Broome N., Phillips A. and Sandwith T. (2014). Governance of Protected Areas: From understanding to action. IUCN Best Practice Protected Area Guidelines Series No.20, Gland, Switzerland. IUCN. xvi + p. 124.

implementing a bottom-up approach. This approach combined the knowledge from local populations and scientific expertise through combined efforts. This included joint surveys, mapping and tools tailored for the issue at hand using a holistic approach such as that implemented through necessity by farmers [CON 86].

For example, in relation to protected areas, the 5th IUCN World Parks Congress in Durban in 2003 emphasized the importance of a participatory approach for the management of protected areas and biodiversity as well as for the assessment of biodiversity [TUC 05, LAW 10].

Participation may cover a specific field such as forest management [LAW 07] or management of biosphere reserves created under the aegis of UNESCO with the involvement of NGOs [RAG 13a] or it could operate on a voluntary basis to assess biodiversity [LAW 07] as advocated by Aichi Target 18. As early as 1999, the Conference of the Parties called for the participatory management of wetlands in Resolution VII.8. The Ramsar Convention Secretariat then proposed a manual on participatory competences in 2010 to clarify participatory environmental management approaches to wetland management.

In terms of cultivated biodiversity, there are some research programs in the Consultative Group on International Agricultural Research (CGIAR), such as the Seeds for Needs program, which is based on participatory selection by farmers of local varieties of seeds.

In the field of biodiversity, as we have just seen, there are many examples of participatory approaches at various scales and on more general or more specific issues.

Box 11.2. *Local governance and participatory approaches*

There are a number of legal tools that enable a convergent study of health and environmental problems at the national level. These may be Environmental Impact Assessments (EIAs), which are internationally recognized but controversial in terms of their implementation (independence of experts in relation to the project, consideration of the time scale, procedure, partial access to information). These differ from one State to another but they do consider the environmental impact and human health of projects, studies or interventions in the natural environment. Nevertheless, it should be noted that this tool still barely integrates biodiversity despite the decisions of various COPs of biodiversity-related conventions[27]. Health impact studies should also be mentioned here; these can be carried out before a project is implemented, in order to study the effects on human health and the environment and alert public authorities on practices that may have an adverse effect (use of pesticides and agricultural production, water pollution). The procedure is still very diverse in different countries. In Thailand, some health impact studies are conducted by local communities and they can challenge public authorities on health issues after conducting their own investigations[28]. Finally, the WHO-supported National Environment Plans provide intersectoral planning and a multidisciplinary approach to health and environmental issues, and clarify the roles and responsibilities of the various stakeholders[29].

The governance of biodiversity and health is not fixed and is the result of various initiatives and networks for which the mission was to manage issues defined to a greater or lesser extent in the field of health and environment or biodiversity. The effectiveness of this governance lies in good connections between the different levels of decision-making but, above all, in the actual desire of States to take control of health and biodiversity issues. By adopting regulations that coordinate the principles to which they have acceded as a party to international agreements, States can decide to show real involvement in the conservation of biodiversity and the protection of public health.

These regulations need to be part of a broader framework for environmental justice, in which citizens and local communities can truly be involved in the decision-making process and particularly access and benefit-sharing of biodiversity. Let us note that many decisions of the Conferences

---

27 Prieur Michel, [PRI 11], "*Instruments internationaux et évaluation environnementale de la biodiversité: enjeux et obstacles*", *Revue juridique de l'environnement*, 5/2011 (special edition), pp. 7–28.

28 Lajaunie C. and Morand S. (2015). A legal tool for participatory methods in land systems science: the Thai model of Health Impact Assessment and the consideration of zoonotic diseases concerns into policies. GLP newsletter, no. 11, Avril 2015, pp. 30–33.

29 These plans exist in Europe (see http://www.who.int/heli/impacts/nehaps/en/, last accessed on 07/05/2017), for France the current plan is the National Health Environment (PNSE3) 2015–2019. These plans are also being implemented in Southeast Asia (see WHO, UNEP, Charter of the Regional Forum on Environment and Health Southeast and East Asian Countries – Framework for Cooperation 9 August 2007, Bangkok, Thailand).

of the Parties to the CBD have given States the responsibility of benefit-sharing with local communities[30]. Fully involving local communities implies that a State has planned for local populations to participate in the biodiversity management process with real support and recognition of biodiversity practices in these communities[31].

Implementation of the Nagoya Protocol on Access and Benefit-Sharing into the national legislation of party States may be an opportunity to consider this concept in all its complexity and richness, and to make the sharing of knowledge effective, in terms of sharing of benefits that arise from the use of traditional knowledge or scientific cooperation[32]. Beyond the responsibility of States, benefit-sharing is also a duty of the scientific community, not just for research with commercial purposes but also for public scientific research.

30 And even before the Nagoya Protocol, see Morgera E. and Tsioumani E. (2010), The Evolution of Benefit Sharing: Linking Biodiversity and Community Livelihoods. Review of European Community and International Environmental Law, 19: 150–173.

31 Morgera E. and Tsioumani E. (2010), pp. 162–163.

32 Morgera E. Fair and Equitable Benefit-Sharing at the Cross-Roads of the Human Right to Science and International Biodiversity Law. Laws 2015, 4, 803–831.

12

Ethics, Values and Responsibilities

12.1. Introduction

This chapter is not intended to be a summary of all the studies that exist on ethics, and even less is it intended to be a crash course on ethics in biodiversity and health. Instead, our aim is to open some avenues for reflection on the ethical implications of research in biodiversity and health. Recent health crises, such as the Ebola epidemic in West Africa, have often led to ethical debates when it comes to health interventions in times of crisis. These interventions tend to rely on forced quarantine or use of new vaccines that have not yet been registered or, for veterinary sanitary crises, on mass slaughter of infected and non-infected domestic or wild animals that are promulgated as a measure to eradicate infectious diseases. On the other hand, calls for "bioethics" [THO 15] or a "global health ethic" [GOL 15] are not necessarily unanimous among ethicists [VER 16].

The development of new technologies, such as gene editing and gene manipulation (CRISPR-Cas9) and the creation of hyper-virulent pathogens (so-called "gain of function" techniques) raise other ethical issues and responsible research challenges when conducted in a situation of uncertainty on the consequences of these new technologies [COU 17, LIP 14].

Ethics and accountability principles apply to all aspects of scientific research and its technological and engineering applications, regardless of whether they are "hard" (based on state-of-the-art technology in the fields of medicine, agronomy or environment) or "soft" (based on local or global governance, health, ecosystems and the environment).

Ethical reflection in the field of biodiversity/health requires first and foremost examining the involved sciences (biology, ecology, evolution, human medicine, animal medicine, anthropology, law, etc.), their epistemology, representations and the need for scientific pluralism, which is essential to establish genuine interdisciplinarity. Once this first point is analyzed, it becomes easier to consider the elements of ethics, values, morals, rights, justice and responsibility, which are called upon in such research. This chapter cannot be exhaustive, and only presents a few frameworks on health ethics, environmental ethics, animal ethics and their implications both in the field work of the researcher and for local or global governance. To conclude, this chapter emphasizes that if biodiversity and health research require recognition of scientific pluralism, then they also imply the acceptance of a pluralism of ethics.

12.2. Pluralism of scientific approaches

Auguste Comte, the father of science history, described what he called the "hierarchy of sciences", whereby the sciences progress through successive stages of development: astronomy first, successively followed by physics, chemistry, biology and, finally, sociology. Thus, the "*hierarchy of the sciences describes not only the complexity of the phenomena studied by different sciences, but also their stage of intellectual development*" [COL 92]. Storer quoted by [COL 92] made the distinction between the so-called "hard" natural sciences and "soft" social sciences. Kuhn [KUH 70] classified the sciences according to the state of development of their paradigm and the sharing of theoretical structures and methodological approaches for which there is a high level of consensus among researchers. In a critical approach to the hierarchy of science, Cole [COL 92] provided a framework of perceptions on science, which are important to consider for an interdisciplinary approach. The sciences do not differ on the reality of different fields and forms of their scientific practices, but on the subjective perceptions of these domains and forms of practice (Table 12.1).

Characteristics	"Top" of the hierarchy	"Bottom" of the hierarchy
1. Development of theory	Highly developed theory; paradigm-driven research; high levels of codification	No theory or weak generalization; pre-paradigmatic phase; low levels of codification
2. Quantification	Ideas expressed in mathematical language	Ideas expressed in words
3. Cognitive consensus	High levels of consensus on theory, methods, importance of problems, importance of individual contribution	Low levels of consensus on theory, methods, importance of problems, importance of individual contribution
4. Prediction	Possibility of using theory to make verifiable predictions	Inability to make verifiable predictions
5. Obsolescence rate	High proportion of references to recent works; rapid obsolescence of previous works; significant accumulation of knowledge	Low proportion of references to recent works; old works remain as important as new ones; low knowledge accumulation
6. Rate of growth	High rate of increase in new knowledge	Low rate of increase in new knowledge

Table 12.1. *Characteristics of the different types of sciences (according to [COL 92]) and their links to the positivism of Auguste Comte and his "hierarchy of sciences"*

Such perceptions of a "hierarchy of sciences" lead to the problem of scientific monism with regard to scientific pluralism. Scientific monism is defined by Kellert *et al.* [KEL 06] as: "*the ultimate aim of a science is to establish a single, complete, and comprehensive account of the natural world (or the part of the world investigated by the science) based on a single set of fundamental principles; the nature of the world is such that it can, at least in principle, be completely described or explained by such an account; there exist, at least in principle, methods of inquiry that if correctly pursued will yield such an account*". Kellert *et al.* [KEL 06] defined scientific pluralism as the multiplicity of approaches that characterize many fields of scientific inquiry. Scientific pluralism is not necessarily deficient in monism and should therefore not be placed at the bottom of the scientific hierarchy. Finally, in Kellert *et al.* [KEL 06], having argued for the importance and recognition of knowledge that is borrowed between scientific and non-scientific fields, Kellert defined disciplinary pluralism as a form of interdisciplinarity and a transversal formation of that which is interdisciplinary. Miller *et al.* [MIL 08] also emphasized that scientific pluralism enables integrated studies to be conducted successfully. Pluralism in studies on the management of socio-ecological systems avoids the predominance of certain types of scientific domains (avoidance of the hierarchy of sciences) and the mere stacking of insufficiently integrated research. However, to achieve this, scientific pluralism requires an understanding of the epistemology and scientific representations of the disciplines being convened (Table 12.2, [MIL 08]).

Knowledge as the study of mechanisms	Knowledge as contingent	Knowledge as narrative
Belief in objectivity, replicability. Knowledge acquired through the scientific method seeks to demonstrate causality and allow prediction.	Importance of agent and context. Knowledge seeks causality; it is based on behavior, variability in relation to social norms.	Interpretive and critical knowledge. Knowledge is inherent to the object and represents values that can be shared or held individually.
Nature as mechanistic	Nature as a complex adaptive system	Nature as constructed

Table 12.2. *Metaphors of scientific knowledge and their implications for studying nature (from [MIL 08])*

To avoid relativism of scientific knowledge, some authors introduced the principle of scientific impartiality [LAC 99]. Impartiality is constitutive to science and must be reflected in cognitive or epistemic criteria of acceptance or rejection of a scientific theory, and not on social, ethical or aesthetic values. Lechopier [LEC 11] pointed out that "*the concept of impartiality, thus analyzed, forms the basis of a realistic epistemology of scientific knowledge: it is possible, by guiding the inquiry into cognitive values, to know the reality, it is independent of any other value judgment*". This proposal was opposed by post-modern philosophers who emphasized the impossibility of developing scientific activity outside the representations that circulate in societies.

Scientific pluralism therefore requires questioning the ethics, and the values, of scientific fields or actions.

12.3. Some definitions

Ethics are usually divided into three branches to which morality is added [HOO 92, JEA 08]:

– meta-ethics theorize on the very nature of ethics, that ethical principles are of a divine order, arising from biological necessities or resulting from rational personal interest. For example, evolutionary ethics is a meta-ethic [DEN 95, SCO 11] as clearly stated by Wilson [WIL 75], who suggested that "*scientists and humanists should consider together the possibility that the time has come for ethics to be removed temporarily from the hands of the philosophers and biologicized*";

– normative ethics seek to determine whether a conduct or action is good or bad according to the theories of obligation, responsibility and fairness;

– applied ethics is a combination of ethical principles and other principles, such as economic, legal or political principles, and factual information (such as scientific information) to develop practical principles for particular situations (vaccination, forestry rights, hunting licenses, conservation of biodiversity, etc.).

Morality refers to the actual practices of a group, a community or a culture that demonstrate and maintain a certain ethic [HOO 92], such as religious morality.

Kernohan [KER 12] presented a classical ethical situation that involves a moral agent (a person like a scientist, a society or government) carrying out an action with consequences for certain beneficiaries (or victims), who are the recipients of the consequences of the action of this moral agent (Table 12.3). Each component of this ethical situation corresponds to a different approach to ethics. We can focus on the agent, on the principles that guide the action or on the benefits and harm of the consequences of the action.

The ethics of virtue date back to philosophers from antiquity. A modern approach to the ethics of virtue can be found in the ethics of "care", which we see in human or animal welfare [JEA 08].

The deontological approach focuses on the principles that guide and evaluate an action based on its compliance with rules that the action must respect. Deontologism is an approach of Kantian origin, according to which an action is morally good if it is done through duty or respect for the law;

this presupposes the objective existence of certain universal moral obligations [JEA 08].

Moral agent	Action	Consequences	Recipient
Character of agent	Principles	Benefit or harm	Ethical implications
Virtuous or vicious	True or false	Good or bad	Moral position
Ethics of virtue	Deontology	Consequentialism	Bioethics, animal ethics, environmental ethics

Table 12.3. *Example of an ethical situation involving a moral agent (a person like a scientist, a company or government) carrying out an action that leads to consequences for certain beneficiaries or victims, who are recipients of the consequences of the action of the moral agent. Considering the agent, his or her action, the consequences or the recipients falls under the umbrella of different ethical approaches (adapted from [KER 12])*

The consequentialist approach emphasizes the consequences of an action and evaluates these consequences according to the benefits or the prejudices they produce. An action is morally good if it produces the best possible consequences [JEA 08, KRA 09]. Contrary to deontologism, consequentialism argues that the value of an action is not intrinsic but depends on its consequence, which varies depending on the context. Thus, the hedonistic utilitarianism of Jeremy Bentham or John Stuart Mill, like the utilitarianism of preferences, aims to maximize the happiness or well-being of individuals. Utilitarianism embraces a second principle: the quantity (or expected quantity) of good we produce is the only thing that matters and we must strive to make this quantity (or expected quantity) as large as possible [KRA 09]. An action is chosen if the total quantity of good obtained, subtracted from the amount of harm, gives a larger sum than any other alternative [KRA 09]. A quantity of harm can be inflicted if the amount of good to come is significant. An example is animal experimentation in the medical field. The suffering and death of the animal experimented upon is acceptable by medical ethics if the result contributes to the improvement of human health (but see below).

In his “Theory of Justice”, John Rawls gave priority to the ethics of law (based on deontologism) over the ethics of goodness (based on consequentialism), stating that “*The principles of right, and so of justice, put limits on which satisfactions have value; they impose restrictions on what are reasonable conceptions of one’s good*” ([RAW 71], quoted by [KRA 09]).

As shown in Figure 12.1, Hooker [HOO 92] summarized the components of the ethics of various standards including justice, rights, ethics and values, and meta-standards and their choices.

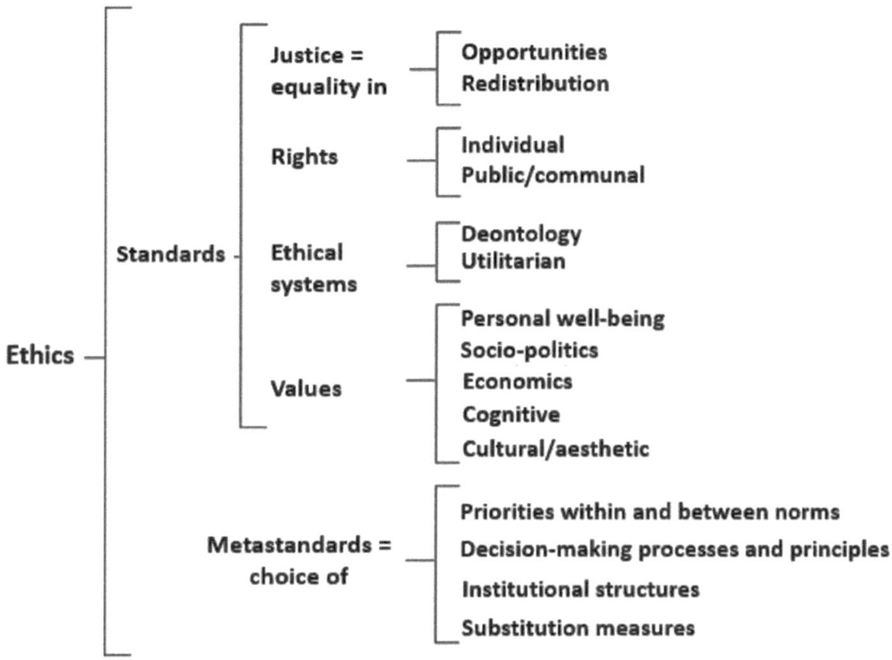

Figure 12.1. *The various components of ethics (according to [HOO 92]). Justice (of opportunity or redistribution), rights (individual or collective), ethical systems (deontology, consequentialism) and values (cultural, aesthetic, cognitive, economic, socio-political and personal well-being) are included among the various standards of ethics. Meta-standards involve choices: priorities within and between standards, decision-making processes and principles, institutional structures and substitution measures*

Kernohan [KER 12] pointed out that each case involves a moral questioning of the consequences of an action on the recipient(s), which raises specific ethical considerations when it comes to livestock and wild animals, or ecosystems. They have specific implications for notions of responsibility towards nature, as described by Passmore ([PAS 80], quoted by [HOO 92]) (Figure 12.2). We have responsibilities towards nature and for ourselves ("for whom?" axis) and these responsibilities are for ourselves, a divinity or nature itself ("to whom?" axis). Why do we have such responsibilities? Because a divinity commands it, because it is in our personal interest or because it is necessary from an ethical stance ("why?" axis). References to divine entities and religions are central to the ethics of health, such as environmental ethics [PET 01] or biodiversity [OBR 10] and its applications in biodiversity conservation with entities such as "Mother Nature" [ROA 03].

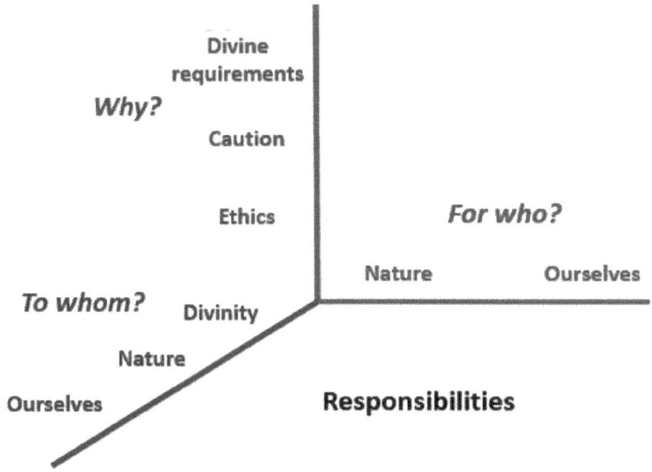

Figure 12.2. *Notions of responsibility towards nature, as described by Passmore (1980, quoted by [HOO 92]) (Figure 12.2). We have responsibilities towards nature and for ourselves ("for whom?" axis) and these responsibilities are for ourselves, a divinity or nature itself ("to whom?" axis), because it is in our personal interest or it is necessary from an ethical stance ("why?" axis). For a color version of the figure, see www.iste.co.uk/morand/biodiversity.zip*

12.4. Humanist and human health ethics

The four moral principles of medical ethics [BEA 94, GIL 94] are: respect for the patient's autonomy and freedom of choice; benefit of the practitioner's action in the best interest of the patient; not be the cause of prejudice or harm; ensure equal justice in the allocation of health resources. The ethics of human health opens into the fields of public health and justice policies and the fields of medical practice and research with bioethics.

The moral principles outlined above originated from the "natural law", the foundation of Christian morality (meta-ethics), that is common to all beings who are capable of thinking and choosing and therefore of having a moral life [BOY 06]. This implies a common morality for all of mankind, embodying responsibilities for all. This universalism implies a consensus on the moral principle and some general standards while recognizing that responsibilities have a global scope [BOY 06].

This moral universalism may come up against the recognition of cultural diversity, as Engelhardt [ENG 06a] underlined in the declaration on bioethics and human rights. Article 10 of the declaration states that: "*The fundamental equality of all human beings in dignity and rights is to be respected so that they are treated justly and equitably*", and Article 12 stipulates That "*The importance of cultural diversity and pluralism should be given due regard. However, such considerations are not to be invoked to infringe upon human dignity, human rights and fundamental freedoms nor upon the principles set out in this Declaration, nor to limit their scope*". According to Engelhardt [ENG 06a], the UNESCO declaration reflects the false assumption of a consensus from morality experts, which should be similar to that of a scientists' consensus. Bioethics is therefore a "field of controversy", a consequence of the existence of moral pluralism. There are significant moral disagreements on substantive issues concerning bioethics policies and health policies, which are not resolved through rational arguments as in the case of scientific controversies [ENG 06b].

In addition to these values, respect of persons must also be considered. Persons must be treated with dignity, as well as with the authenticity and honesty of the action, through the notion of "informed consent". Informed consent is based on the moral principles set out above [FAD 86]. It is transcribed into legislation and requires the patient to be informed in order to consent to a proposed therapy, which is essential in the case of therapeutic trials. Informed consent is mandatory for epidemiological studies, such as monitoring of infectious diseases. In the latter case, obtaining the informed consent of individuals presupposes that they have the knowledge to understand the objectives of the research being conducted and the benefits and potential harms, which will be withdrawn individually or collectively. Informed consent questions the ethics of anthropological research.

12.5. Animal and animal health ethics

The ethics of animal health is different from human ethics in that an animal is not regarded as a moral being but as a sensitive being to which utilitarian principles apply. These principles date back to Jeremy Bentham and, prior to him, Jean-Jacques Rousseau, who both stressed the duties of humans towards animals, as Jeangène Vilmer [JEA 08] reminded us. Animal mistreatment is morally wrong, not because the animal is a reasonable being but because it is a sensitive being. Animal ethics considers animal suffering and death as harm that must be avoided or minimized (in breeding, research, etc.). Animal ethics is very active in the field of animal welfare, which aims to promote the adoption of practices that ensure welfare conditions for domestic and laboratory animals. Animal ethics are based on research in ethology and stress physiology. Animal ethics has led to the setting up of ethics committees for research involving laboratory animals and wild animals.

Animal ethics concern sentient living beings and make animal suffering its main concern [SIN 75]. They pose the question of speciesism, and therefore of anthropocentrism, of animals as a moral subject (as in the case of chimpanzee primates or gorillas) or as an individual "subject of a life" [REG 03]. Animal ethics question new biotechnological practices, such as transgenics and gene editing, not only on the basis of animal welfare principles (animal bioethics) or environmental safety (the precautionary principle), but also on the basis of recognition of the status of a moral subject or animal subject to its own life (which may lead to giving rights to animals) [REG 03, DON 11].

Veterinary ethics is different from animal ethics due to the fact that the veterinarian interacts not only with the animal but also with the client (farmer, private individual and company) and with society (public health, agriculture and trade). A veterinarian can thus find himself or herself in situations where he or she does not provide care but where he or she advocates widespread slaughter, as in the management of infectious epidemics and sanitary crises affecting livestock farms. The development of informed consent in veterinary ethics is thus more concerned with the interactions of veterinary researchers with breeders or any other group of people than with the animals themselves, which are the property of owners. Those animals are taken into consideration by animal ethics.

Animal law deals with the legal status of domestic and wild animals and maintains important interactions with animal ethics, animal welfare and humanistic ethics [SMU 12]. The development of environmental ethics addresses new questions about non-sensitive living entities, such as plants, or complex entities such as ecosystems.

12.6. Environmental ethics

International law has evolved to recognize the right of humans to a decent, healthy and sustainable environment (Chapter 11). Thus, the appeal adopted in 2007 at a conference of non-governmental organizations of senior United Nations officials and the French Head of State highlighted that: "*To promote environmental ethics, we are calling for the adoption of a Universal Declaration of Environmental Rights and Duties. This common charter will ensure that present and future generations have a new human right to a sound and well-preserved environment*" (Paris Appeal 2007, quoted by [TAY 08]).

The Charter raises the question of a balance that must be found between rights, duties and responsibilities. It also raises a question on the place of anthropocentrism, which is inherent in human rights, in the environment. For some commentators, the very existence of human rights in the environment reinforces the idea that the environment and natural resources only exist for human purposes and have no intrinsic value and ultimately create a hierarchy where humanity holds a superior position over other members of the natural community [TAY 08, MAR 10a]. This vision is present in many of the studies cited in previous chapters, where the state of the environment is determined by human needs (living standards, natural resources and health) and not by the needs of other species or the importance and ecological significance of complex interactions between species and the environment. These contradictions of needs reflect the contradictions of scientific and non-scientific representations of nature (Table 12.2).

These different perceptions of nature can also be seen in environmental engineering, as in the control of infectious diseases, and in its representations of nature. In "The Control of Nature", John McPhee (cited by [ROA 03]) mentioned a documentary on the United States Army Corps of Engineers where they presented their fight against "Mother Nature" for flood control along the Mississippi River through management of the river's course: "*This nation has a large and powerful adversary. Our opponent could cause the United States to lose nearly all of her seaborne commerce, to lose her standing as first among the trading nations... We are fighting Mother*

Nature... It's a battle we have to fight day by day, year by year; the health of our economy depends on victory". Environmental action is confronted with ambivalence towards nature and this image of Mother Nature in the form of three representations, as summarized by Roach [ROA 03]: "*Nature is our sweet mother; she loves us, and we her. Nature is our treacherous enemy; she needs taming, now. Nature is our wounded victim; we must heal her who has given us life*".

Rolston [ROL 92] defined an environmental ethic on the principles of respect for plants and animals, endangered species and ecosystems. Thus, contrary to humanistic ethics, which considers ecosystems based on their contributions to human well-being, environmental ethics considers ecosystems to have a broader value by contributing to human well-being, animal experience and plant life. Thus, it "*preserves the integrity, stability and beauty of the biotic community*" [ROL 92].

It would not be possible in this chapter to cover the entire history of philosophy and environmental ethics (see [NAS 89, PET 01, KER 12]), which includes everything from ecofeminism and the ethics of care [WAR 97], deep ecology [NAE 89], Marxist ecology [FOS 10, MOO 15], evolutionary ethics [SCO 11], biophilia [WIL 84] and bioregionalism ([MCG 99], Berg in [GLO 15]) to the most recent debates [ROL 12, CAL 13] (Table 12.4).

The point of view is an important aspect, whether this be individual or collective (community). Animal ethicists often take the individual animal as a point of view for their moral status, whereas environmental ethicists emphasize the moral status of groups or ecosystems as the basis of ethical obligations towards individuals [VER 16] (Table 12.4).

Point of view	Anthropocentrism	Ecocentrism
Individual	Liberal ecologism (market economy) (see [FOS 10]), rights of nature (see [NAS 89])	Animal Rights [REG 03, DON 11]
Community	Cultural diversity [MAF 05], ecofeminism [WAR 97], Marxist ecology [FOS 10], justice [LOW 98], rights of indigenous peoples (see [LOW 99])	Earth ethics [LEO 49, CAL 13], biophilia [WIL 84], deep ecology [NAE 89], bioregionalism [GLO 15]

Table 12.4. *The categories of environmental ethics (adapted from [HOO 92]). These categories and points of view must be linked to the representations of Nature (Table 12.2)*

12.7. Applied and global environmental ethics

Environmental ethicists advocate the need for change in consciousness, attitudes, thoughts, models, beliefs and world view. Their environmental ideas are borrowed from Asian American and American Indian philosophical and religious traditions and can be adopted as a therapy for global diseases [CAL 89]. The ethics of the planet as a "medicine" for the planet was already present in Leopold [LEO 49]. Neither individual changes nor a purely scientific approach is enough to guide the environmental crisis [PAA 05]. The concept of stewardship emerged from environmental ethics as a principle for managing the care of the environment as opposed to selfish exploitation [ATT 99]. Stewardship postulates that humans are a sort of shop steward, tenant, conservator or tutor, which implies a level of concern for the present and for the future (see also the concept of "ecological solidarity", [MAT 12]). Good stewardship requires justice, truth, sensitivity and compassion.

Planetary stewardship or "Earth stewardship" requires integrating scientific and traditional ecological knowledge. Biocultural diversity is not a collection of biological, physical or cultural objects [MAF 05]. Incorporating scientific knowledge and traditional ecological knowledge is a dynamic process, sometimes a source of conflict, but requiring intercultural dialogue and negotiation [ROZ 15]. It therefore requires deconstructing and reconstructing the representations of nature to learn from ecologies of others [DES 05, DES 11]. Earth stewardship must integrate intercultural, interdisciplinary, inter-institutional and international processes that generate forms of co-management of ecosystems and thus avoid the "Tragedy of the Commons" [OST 90]. The concept of "global public goods" has been embodied in the maintenance of the integrity of ecological systems as emerging ecological processes of a complex system (Crabbé and Manno).

According to Razzi *et al.* [ROZ 15], the integration of ethics and ecological sciences must occur through environmental citizenship. This view of the environmental ethic of planetary stewardship (Figure 12.3) explicitly refers to non-Western philosophies, and Amerindian ones in this case, for the conceptual framework of the IPBES [DIA 15a, DIA 15b], even if the IPBES ecosystem services' approach has been criticized by environmental ethicists [MAR 14].

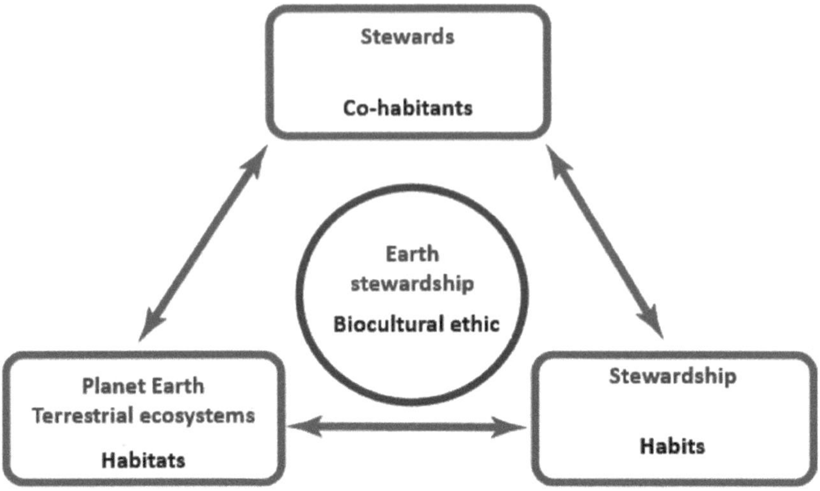

Figure 12.3. *Biocultural ethics are connected through three essential components of Earth stewardship: habitats and planet Earth, stewardship and practices, stewards and cohabitants (from [ROZ 15]). For a color version of the figure, see www.iste.co.uk/morand/biodiversity.zip*

The vision of global governance raises other ethical and legal conflicts. The Millennium Development Goals, 2000 and the United Nations Sustainable Development Goals [UNI 15] have global objectives, including health, the environment and biodiversity, but universal multilateralism can be rejected by nation states [MAC 08a]. A cosmopolitan environmental ethic with the notion of environmental citizenship may prove to be contradictory with the loyalty that citizens owe to their nation-state. Environmental advisers must co-exist within their government with national defense advisors who often view foreign policy as a continuous state of "preparedness for war" ([DOW 98], cited by [MAC 08a]).

12.8. Ethics of foresight and scenarios

Mouquet *et al.* [MOU 15] addressed ethical aspects of predictive ecology, such as the acquisition of data and knowledge of people, local communities and state services, conditions for access to them, transparency of mathematical models and scenarios, dissemination of scientific results, and results and limitations of predictions and projections. Many scientific studies on modeling, projections and scenarios are used to negotiate issues long before any consensus has been reached within the scientific community. Although the "culture" of uncertainty is deeply rooted in the scientific community, it is still difficult to communicate (and sometimes poorly communicated) to the general public, concerned communities, managers and politicians. Scientific controversies are numerous in ecology, such as loss of biodiversity in the face of climate change, the impacts of invasive species and the link between biodiversity and emerging diseases. They are also unclear and some scientists use social media and networks to shift their views instead, rather than shed light on the origins of controversies and their implications.

According to Mouquet *et al.* [MOU 15], there are two distinct aspects involved in the production of scenarios: the ethics of scientific process that generates predictions and the ethics of using predictions in the public debate and governance. Scientists are therefore responsible for communicating the limitations and uncertainties that are associated with their predictions in order to avoid generalizations and unjustified use. These require communicating about consensus and uncertainty.

Another issue is the social representations that accompany the scientific debate [PIE 07], as these have a causal impact on reality by changing the behavior of those who adopt them. The implementation of scenarios or participatory management for local governance uses participatory methods and therefore the ethics of their implementation, their use and their consequences on social and political life [MAN 07].

12.9. Confronting the ethics network

The confrontation of researchers to different ethics is based on the works of anthropologists Laine [DEL 00] and Newing [NEW 11]. They are applied to research in the field of biodiversity health, or "One Health", and highlight the responsibilities of a student or advanced researcher in the face of ethics and deontologies that are sometimes contradictory in the fields of anthropology, human health, animal health and biodiversity (Figure 12.4).

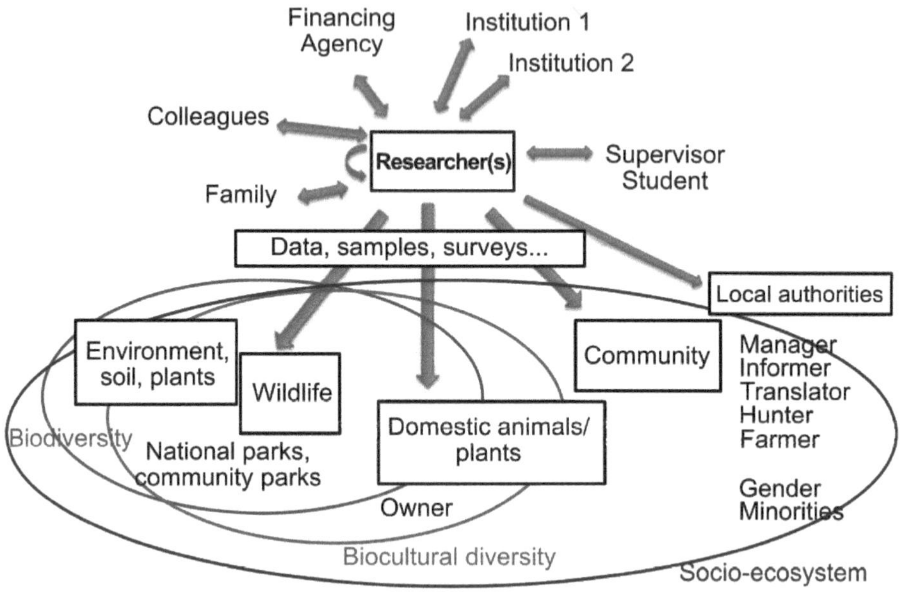

Figure 12.4. *The network of ethics and responsibilities in health/biodiversity research and "One Health" (see text). For a color version of the figure, see www.iste.co.uk/morand/biodiversity.zip*

It all begins with a written research project proposal that has to be validated by a research institution (or even several institutions if the study takes place outside the researcher's country), legal authorities (national parks, ministries), the supervisor if it is for a student and a funding agency (or student grant). This project and its research permits are complemented by various documents from ethical committees for human and animal health along with informed consent requests from those who will participate in studies on their health or the health of their animals. Studies on wild animals (and also domestic animals) also require validation of a protocol for handling animals, which must meet welfare criteria and legal issues (wildlife on the CITES list). The field researcher therefore has responsibilities towards all these institutions and agencies, but also towards his or her supervisor for a student, to his or her students for a supervisor, and colleagues or others who may eventually join the study, family and himself or herself, including elements concerning safety and health of all those involved. Arrival on the ground requires permits and letters of introduction to local authorities (administrative, health, etc.) and community leaders.

On site, researchers collect different types of elements, not only from surveys and interview data to biological samples from humans (stools, blood samples), domestic or wild animals (and their microbes and parasites), but also possibly from the environment and plants. It is therefore necessary to obtain liberally given, free and prior informed consent before starting any investigation with a person. Signed consent must be acquired in the field of human health or domestic animal health. However, within the context of anthropological interviews (qualitative analysis), prior signed consent may create suspicion and concern to the participant, so it may be preferable to just get oral consent (which implies that researchers inform the participant of the purpose and use and level of confidence of the information that will be obtained). Access to animals, whether wild or domestic, private areas (home, fields) or communal places (forests and community spaces) must also be authorized. Thus, if a domestic animal belongs to its owner, a wild animal is never totally free and access to it may depend on a hunter, the community or an institution (nature reserve). Compensation and payment arrangements must be set up for research activities like translators, informants, breeders who will provide access to their animals, use of equipment and local health workers or veterinarians.

In the field, the behaviors of researchers can have significant consequences, such as the death of an animal during handling, and their attitudes may be shocking to communities or local authorities. Researchers may be confronted with situations of significant moral dilemma, such as witnessing ill-treatment of weak people and animals, trafficking of wildlife or other elements of biodiversity, and aspects of corruption. Researchers then gain responsibilities in light of the answers obtained to their questions. Elements may relate to aspects of illegal practices, or highlight minorities (women, ethnic communities) for aspects of health (human or public) or practices. The ethics of anthropological research may conflict with the ethics of human health, for example, in terms of publishing the names of the villages studied. It is common for biomedical studies to mention the exact names of villages, with their precise geographical positions, while anthropological studies usually mask the names of these villages (not to mention the ethical impossibility of giving the precise geographical location).

The knowledge associated with biological samples, whatever they may be (microbes, plants, animals) cannot be dissociated from the communities to which they belong. Using this knowledge for commercial operations concerning living matter (patents, etc.) without local communities benefiting from them are acts of biopiracy (not abiding by the laws from the Nagoya Protocol).

Numerous qualitative and quantitative data are obtained. Who owns these data? What level of accessibility is there, and for whom, between "open data" and the need to respect confidence and knowledge? Biological samples are collected. What level of safety (personal, collective, animal welfare, environmental), contamination and spread are ensured during the collection and transport phases? Where will samples be stored and with what level of security? Who can access them and under what conditions? How are the associated metadata (geolocation, knowledge and practices) preserved and who has access to them and under what conditions/restrictions?

Finally, an important aspect is the feedback of scientific work to agencies and institutions, colleagues with questions about the knowledge and modes of communication (such as geo-positioning of communities, sensitivity of elements obtained). Feedback to communities and local authorities is sometimes neglected and is therefore an unethical element of scientific work. This feedback paves the way for communicating scientific knowledge but above all, it provides a place for scientific and non-scientific knowledge to meet, as well as the associated representations. Some results can be used in certain aspects of local or more global governance, or for power stakes for groups or individuals for which the consequences for communities may be quite different from what could have been imagined at the beginning of the study.

12.10. Necessity of pluralism of ethics

As Hooker [HOO 92] noted, every aspect of life is filled with tensions and these tensions also concern ethics. Let us take the example of the avian flu crisis in 2017 in Southern France. Bird slaughtering measures were taken, including infected and non-infected animals, in order to eradicate the presence of the virus on the territory. These measures were taken because of public health and veterinary health principles, which also had economic implications following the banning of poultry exports from France. The avian flu crisis in the South-West had consequences for the poultry market in North-West France. We could assume that the slaughtering took place to follow ethical recommendations on animal welfare, but not animal ethics and even animal rights. It is likely that the consequences of slaughter on the psychological health of pastoralists and "slaughterers" have not been fully considered. Finally, the proposed slaughter of some local duck breeds was at odds with the Aichi objectives from the Convention on Biological Diversity, and in particular Target 13: "*By 2020, the genetic diversity of cultivated plants and farmed and domesticated animals and of wild relatives, including other socio-economically as well as culturally valuable species, is maintained, and strategies have been developed and implemented for minimizing genetic erosion and safeguarding their genetic diversity*". International organizations such as the FAO may have contradictory opinions with, on the one hand, biosafety measures, which lead to a reduction in genetic diversity of livestock and, on the other hand, defense of cultivated genetic resources and the associated knowledge [FAO 15].

How scientific results are communicated and how they are interpreted by the media or health authorities can have important consequences. For example, reading that wild birds (in the case of H5N8 avian influenza outbreaks in Europe) or bats (in the case of Ebola outbreaks) are the "keys" to these epidemics can lead to confusion among health crisis stakeholders and have effects on wildlife (introduction of slaughtering measures, for example). Indeed, if wild birds or bats are the reservoirs of viral diversity and therefore the biological origin of agents, we must consider the socio-ecosystem as the ecological origin of transmission, with the crisis originating in the socio-economy. Wild birds and bats are only slightly indebted for economic losses and more dramatically for the deaths caused.

These examples demonstrate the importance of values borne by different scientific disciplines and the various actors. The fields of public, animal and environmental health research, and in particular epidemiology (and its various branches), determine the choice of variables, parameters and factors that are considered to be relevant enough to study and which are the subject of strategic choices [LAC 99]. The so-called "modern epidemiological" approach often focuses on risk factors associated with individual genetic or behavioral characteristics. This approach makes it possible to apply targeted action to correct individual factors. It leads to individual medicine and bioethics research. The so-called "eco-social" approach focuses on global, socio-economic and environmental factors, which are the contextual causes of the considered health problem. These determinants will be acted upon through systemic action, which requires ethical pluralism.

12.11. Conclusion

The quality of goods and services that ecosystems provide to human societies depends on their being in good condition and functioning (Chapter 9); this implies an analogy of "good health" in ecosystems. Proper functioning of the planet's ecosystems requires continuous "adaptation" of modern societies through governance (Chapter 11) and stewardship, such that the level of well-being and health of individuals is at its maximum while maintaining an optimal level of biodiversity and biocultural diversity. Such an objective requires taking into account values that are conferred on human health, non-human health and nature.

Do our globalized societies need global health ethics? Such ethics cannot be limited to bioethics and the deontological ethics of science, but must be open to animal ethics and ethics of the environment. It must take into account the plurality of conceptions of nature and living beings. It must reflect on the need for justice of populations and societies that are most exposed to environmental and health crises in a world that has been globalized by a liberal economy. It must consider individualist and collectivist values. It must be adaptive, while being pragmatic and normative, integrating the needs for development and well-being as well as the conservation of biodiversity.

The cultural diversity of representations of nature and health questions the "ecosystem services" that scientists seek to predict and implement in governance tools, such as payment schemes for ecosystem services [CAL 03]. A "global ethic" for health may not be more essential than "One Bioethics" (from "One Health") would be. As Verweij and Bovenkerk [VER 16] pointed out, such ethics are more the domain of meta-ethics that corresponds to a moral belief in "health" and "planetary health".

13

The Role of Law, Justice and Scientific Knowledge in Health and Biodiversity

13.1. Introduction

As we have seen, there are many definitions of health and even the One Health approach calls on us to abandon a strictly anthropocentric vision of health in order to consider that health and human well-being depend on the health of all living things and that of the environment in general. Far from being restrictive, this approach calls for a much broader approach to health than just human health.

Similarly, biodiversity implies diversity, whether it be cultural diversity as we saw in Chapter 2 or, more generally, diversity of approaches to scientific questions. The complexity of interactions between biodiversity and health such as those presented in this book invites us to use a variety of scientific disciplines to describe reality through distinct methods. Interdisciplinarity, which is a reflection of this diversity, is *de facto* imperative when examining the links between biodiversity and health.

Yet when it comes to reporting on the state of scientific knowledge, notably to inform political decision-makers or elaborate on a rule of law, researchers find themselves faced with a paradox, which Charadeau [CHA 10] presented as follows:

> "At a time when there is much talk about multidisciplinarity or interdisciplinarity [...], the institutional system and a certain corporation of researchers only take monodisciplinary technicist studies into consideration in the name of scientificity".

This raises the question of taking complexity into account in the presentation of scientific knowledge. From the outset of knowledge production, this approach also implies a proactive approach from the researcher to escape scientific reductionism by assuming complexity, considering diversity – notably disciplinary – as an integral part of research and by linking knowledge to it. E. Morin [MOR 17a] highlighted that:

> "Dispersion and compartmentalization of knowledge in specialized disciplines elucidate the major problems that arise when one associates the knowledge encompassed within the disciplines. Therefore, essential questions are eliminated. Ignoring them maintains an ignorance that reigns not only on our contemporaries but also on savants and experts who are ignorant of their ignorance".

13.2. Complexity, scientific knowledge and informing political decisions

In the 1990s, Funtowicz and Ravetz [FUN 93] questioned the role of scientific research in informing policy, in particular regarding environmental and health challenges, sustainable development and risk assessment issues. This study raised the question of the reliability, validity and objectivity of scientific results on which political and legal decisions are based. With the theory of post-normal science, the authors proposed to manage complex scientific problems associated with political decisions that were linked to important and controversial issues in a new way. They proposed to integrate the different degrees of uncertainty that accompanies given scientific information into the debate, reflecting the complexity of reality and presenting these uncertainties in a transparent way to the various actors involved in the political decision-making process. This implies an assessment of the quality of information at the decision-making source, in other words an assessment of the robustness of scientific evidence and its appropriateness in the political decision at stake. The authors also considered that the information used to solve these problems must not only emanate from scientific circles but may come from a multitude of actors who are involved in the political debate. They advocate taking multiple perspectives and skills and knowledge that are broader than those of the scientific community itself into account, and thus a peer community that includes non-scientific actors who are involved in decision-making and, subsequently, they advocate a participatory democratic approach.

This reflection is also part of the literature on wicked problems (which differs from complex problems, see [MOR 13a]) and is linked to public policies by Rittel and Webber [RIT 73], who suggested that wicked problems do not have objectively good solutions that can always be applied successfully. In fact, this type of problem cannot be solved using a classical scientific approach because it calls for an answer to social problems, which involves a variety of behaviors, many people and different sectors: each context is unique and each problem is unique. These problems are socially complex and the solutions imply coordinated action of a set of actors [AUS 07]: individuals, non-governmental organizations, private companies, government institutions from the local to international scale.

This also raises the question of neutrality and the use of scientific "evidence" by policy makers or in the definition of rules of law: knowledge as it is given to institutional leaders can erase disparities [TUR 16] and, as such, the role of discourse, meaning the angle chosen to present the information, should not be ignored.

About the question of neutrality, in the context of rapid and global changes, we must also question the degree of uncertainty. Scientific uncertainty is often associated with risk and leads to expert opinion with compromises. As proposed by A. Stirling [STI 10], pluralistic and conditional scientific advice allows political authorities to know the different options and allows a more sophisticated political debate by understanding the extent of what is unknown. This also means respecting the principles of ethics and scientific integrity.

First, let us consider what legal issues are underway for the development of a law that is in line with tangible issues. Next, we will examine citizens' use of scientific knowledge for the purposes of collective justice. Finally, we will discuss how scientific knowledge can be viewed from the perspective of human rights.

13.3. For a law that is in line with reality: difficulty in implementing the principles of transparency, accountability and participation

Since the 1990s, at the international level, the principles of transparency, accountability and State responsibility have been considered in relation to the various fields of action of States (economic, environmental, health, social) as elements that are necessary for good governance and to achieve the rule of law, particularly through development organizations. They are closely linked to the participation of citizens in political decision-making and the issue of their representation.

These principles are contained in the Sustainable Development Goal no. 16: "Promote just, peaceful and inclusive societies", which are further developed in Goals 16.6 and 16.7:

> "16.6 Develop effective, accountable and transparent institutions at all levels;
>
> 16.7 Ensure responsive, inclusive, participatory and representative decision-making at all levels".

The problem lies in the implementation of these rather general principles and the way that they are translated by different stakeholders: they can be development actors (donors), government institutions or defenders of human rights and rights of local communities. The main apprehension is that these principles, which leave a wide margin of interpretation, are applied superficially to respond to a technocratic approach but without real practical translation on the ground [CAR 14]. Thus, the principles of participation and empowerment can be applied with varying degrees of real implementation (for typology of modes of participation, see [AGA 01]). As Mohan [MOH 07] pointed out, participatory development must take into account the different political spaces (at different spatial scales), in which participation can be effective. Responsibility and accountability should not only be sought at the local or national level but should more broadly involve networks of actors (institutions, civil society, non-governmental organizations), which are placed at different levels of the decision-making process (from global to local). It also examines the diverse forms of political communities in which poor or marginalized populations protect their livelihoods and well-being.

These principles of transparency presuppose that citizens can participate upstream in political decisions of States or local institutions, and that they can also be held accountable for the political decisions made, management of public affairs and their consequences on the environment and health.

To apply environmental law, we must consider environmental issues *inter alia* within the national legal system and sensitize magistrates on environmental issues and their connection to health. Having adopted a strategy for 2020, the Asian Development Bank considers the environment and climate change as two crucial areas of action and, in 2010, organized a Symposium of Asian Judges on Environmental Decision-Making, the Rule of Law and Environmental Justice that was in line with the principles asserted in 2002 at the Johannesburg Summit.

This Symposium recognized the need to strengthen the training of environmental lawyers, who could act in favor of a more respectful state of law for the environment. The information sharing and experience on good practices in similar cases in the region of Asia were widely acclaimed [ADB 11].

The possibility of demanding accountability required improving citizens' access to justice (Principle 10 of the 1992 Rio Declaration) or alternative forms of conflict resolution.

The number of tribunals or courts that specialize in environmental issues has increased dramatically over the last fifteen years in response to citizen demand for environmental justice across the world.

To facilitate access to these tribunals, the United Nations Environment Program proposed to implement various mechanisms, such as itinerant tribunals that meet people, provides free translators and procedural aids (UNEP, see [PRI 16]).

It should be noted that under the auspices of the United Nations Environment Program in 2016, the Second United Nations Environment Assembly placed environmental justice at the heart of action to promote a healthy environment. This took place just after the First World Congress of Environmental Law organized by the IUCN, which gave rise to a declaration from the participants on the rule of law in the field of environment.

States recognized the importance of affirming the rule of law (with strong institutions and the leading role of magistrates) in achieving genuine environmental justice. This statement emphasizes the universal responsibility of citizens, States and private entities in considering and promoting the welfare of other species and ecosystems (Principle 1: Responsibility to Nature). Nature is considered here as a common heritage of humanity (Principle 2: Right to Nature and Law of Nature). The Declaration recognized that effective implementation of law is essential and proposes a number of tools, including considering the customary rights of local communities.

Resolution 2/5 of the Second United Nations Environment Assembly on the implementation of the Sustainable Development Program to 2030 recalled the need to strengthen the science–policy interface for the implementation of the environmental component of this program. It called for the use of information from a wide variety of sources and to develop cooperation with partners other than those of the environmental community, drawing on tools, assessments, expert groups and national and international network information systems.

13.4. Scientific knowledge used by citizens for environmental justice

Scientific knowledge can also be used simultaneously for a science-society dialogue (see Chapter 10), but also for better environmental justice.

As the Executive Director of UNEP noted in his inaugural speech at the first IUCN World Congress on Environmental Law in 2016, the number of environmental disputes is increasing considerably, as are international texts and constitutions that refer to the concept of a healthy environment.

Over the past ten years, there has been strong demand for new forms of governance that allow decisions on environmental issues to be fair, fast and inexpensive [PRI 10]. This demand was the result of both awareness and concern from citizens and non-governmental organizations around the world.

A 2017 study (UNEP 2017) provided an overview on global climate change disputes: since the Paris Agreement made an inventory of existing national commitments to keeping global warming below 2°C, the commitments served as a basis for complainants who wanted to highlight the inadequacy of efforts being undertaken by national governments to combat climate change and its impacts, and to protect individual rights.

Sustainable Development Goal no. 13: "Take urgent action to combat climate change and its impacts" includes the following goals:

> "13.2 Integrate climate change measures into national policies, strategies, and planning
>
> 13.3 Improve education, awareness raising and human and institutional capacity on climate change mitigation, adaptation, impact reduction, and early warning".

Complainants therefore use scientific and technical knowledge on the evolution of climate change and base themselves on forecasts for temperature trends and climate variability (see Chapter 10), which are becoming increasingly reliable. The complaints relate to the lack of government efforts to adapt to climate change, particularly in countries where climate change and ecological disasters lead to forced migration, which is an increasing phenomenon [COU 07]. Climate change is presented as a super-wicked problem because of the multitude of socio-ecological interactions involved and because of changes over long periods of time with different effects, depending on the geographical location [LAZ 09]. In this sense, the climate problem is very representative of the difficulty in understanding environmental changes and their effects, and hence the difficulty of establishing clear causal links and accountability. These characteristics turn climate change and environmental problems in general into complex issues that need to be considered in the law and therefore brought before the courts.

This civic commitment for better environmental justice requires improved access to scientific information and environmental awareness [GEL 16]. This type of public interest litigation aims to protect the national environment and the collective socio-environmental benefits. It has been developing rapidly in India [AMI 13, GIL 16], where it targets climate change in cities [BEE 16] and in China with the adoption of a new law on environmental litigation [UNI 15b, LIN 15a].

The multiplication of environmental litigation in the public interest arises from voluntary participation, through commitment from citizens and not from any decreed participation. To facilitate this type of litigation, protective measures have been proposed to counter strategic lawsuits against public participation (SLAPP, see [PRI 89]). These prosecutions are intimidation techniques aimed at hindering political participation of citizens or organizations in a public setting. The existence of these techniques must therefore be taken into account in order to counteract them through rejection procedures and protect complainants who act in favor of the environment (UNEP, see [PRI 16]) and to defend their fundamental rights.

13.5. Human rights and the right to science? Environmental and health challenges

One of the specificities of environmental law is that it relies heavily on science. It is based on an understanding of ecological and environmental conditions and their systemic evolution when faced with global changes. This is what led Robinson [ROB 03] to suggest that a real partnership exists between environmental law and science.

The use of scientific knowledge by citizens to protect the environment and health depends, as we have seen, on access to this knowledge, which must be as open as possible. This issue of access raises the issue of recognizing the right to science as a human right.

It is worth noting that a report by the Human Rights Council to the United Nations General Assembly in 2017 highlighted the negative impacts of the use of pesticides on human health, the environment and society. This report, which was drafted by the Special Rapporteur on the right to food, highlighted that adverse effects of pesticides, through their impacts on health and the environment, undermine the right to adequate food (Article 11 of the International Covenant on Economic, Social and Cultural Rights 1966) and the right to health (Article 12 of the 1966 International Covenant) for present and future generations. It may be added that they also infringe upon the right to a healthy environment as affirmed by the 1972 Stockholm Declarations and the 1992 Rio Declaration, and is also recognized in many constitutions across the world, although the report does not explicitly mention this.

Moreover, this report completely re-questions the assertion from the agrochemical industry that pesticides are necessary to ensure food safety and qualifies it as “inaccurate as it is dangerously misleading”. Noting that the most affected populations are subsistence farmers, this report promotes agro-ecological agriculture, which is the foundation of sustainable agriculture.

The recommendations in this report highlight the importance of informing the general public on the health and environmental impacts of pesticides, including the requirement to label food and beverages with the pesticides used and the level of residues.

This example demonstrates the importance of public information and the need to provide access to scientific knowledge, despite all its controversy and uncertainty. The Universal Declaration of Human Rights (Article 27, paragraph 1) already contained the right to science, which was affirmed by Article 15 of the 1966 International Covenant on Economic, Social and Cultural Rights as the right to “benefit from scientific progress and its applications”.

Although the interpretation of this right and its implementation gave rise to a debate that needs to be further explored [SHA 10, CHA 13, MAN 15], especially on the definition of the term “science”, let us note the benefits that scientific progress includes, aside from its conservation and development: the spread of science. The spread of science is an important element of social justice [WEI 13a]. This diffusion is also linked to scientific training and the free movement of people and ideas.

Ultimately, we must connect this right to science with the need to develop a new type of environmental law that takes multiple interactions into account at different spatio-temporal scales, a law that is adaptive and iterative, and that integrates the most recent scientific knowledge. This adaptive nature was also highlighted in the 2016 IUCN Declaration on the environmental rule of law (Principle 13: Progression).

The relationship between the law, justice and scientific knowledge in relation to health, environmental and biodiversity issues is at the heart of social progress. Human rights go beyond a purely legal approach and as Sen proposed in “The Idea of Justice” [SEN 10], human rights are ethical demands that are only raised to the level of human rights after a public hearing, a debate in full fairness. Thus, open engagement in the public debate is central to the very idea of justice.

Conclusion

Towards a socio-ecology of health

The first chapters of this book explored biological and ecological factors that interact with health and well-being. They highlight the importance and complexity of the physiological, genetic, psychological and evolutionary relationships which we have with animals and plants, microbes and parasites, the environment and therefore biodiversity in general. The biodiversity crisis and climate disruption are the expression of an increase in the planet's human ascendancy (the Anthropocene era) [CEB 15, PEC 17]. New environments are characterized by growing homogenization and simplification of natural biodiversity, and also cultural and bio-cultural diversity. Agronomic intensification and the development of industrial livestock production contribute to increasing the appropriation of biological productivity through a certain type of development, to the detriment of natural environments, coupled with an increase in the use of agricultural inputs and biocides (pesticides, antibiotics).

While the burden of infectious and parasitic diseases has decreased overall in the last century, these environmental changes have been accompanied by new infectious diseases that affect humans, wildlife and domestic animals. At the same time, there has been an increase in non-communicable diseases, some of which are directly associated with loss of biodiversity.

In a context of globalization of economic trade and new connections between the local and global levels, research efforts must be oriented towards a socio-ecology of health. It requires a socio-ecology that is rooted in understanding the ecological dynamics of biodiversity and human societies, taking the biological, ecological and evolutionary links between humans and non-humans into account, including parasites and microbes. This socio-ecology of health can contribute to public health systems by maintaining the environmental conditions necessary for the good health of ecosystems, which are necessary for the development of our social ties and our individual well-being.

Towards scientific pluralism

Focusing on the connections between biodiversity and health means acknowledging the necessary contribution of many scientific fields. The call for multidisciplinary work can be only achieved through accepting scientific pluralism and pluralism of methodologies. It cannot be limited to scientific knowledge and practices alone, but must also integrate traditional ecological knowledge. This scientific pluralism, as we have highlighted, also requires recognizing the pluralism of ethics and values. Giving privilege to reductionist approaches or holistic approaches, techno-sciences or socio-environmental engineering is accompanied by a set of values on science, societies or even nature that are often different and sometimes conflicting. Scientific pluralism imposes a debate on its intrinsic approaches, practices, ethics and values [SPA 15, CHA 16, CCN 17].

Towards an adaptive law

The strong interactions between health and biodiversity push us to rethink environmental law in general and biodiversity in particular, taking into account all the factors that may have an impact on biodiversity in order to determine the level of decision-making and the type of regulation that is best able to conserve and protect biodiversity and the health of the planet. Environmental law is responding to the urgency of situations through *ad hoc* solutions, "tinkered" answers. However, it is time to take a step back from these solutions and build a new law that breaches with traditional legal thinking that does not integrate the complexity of environmental issues.

Just as the One Health approach was presented as a necessary paradigm shift to address infectious diseases in an integrated, interdisciplinary and intersectoral way, the time has come to shift perspectives to develop a reflexive and adaptive environmental law that functions iteratively and integrates new scientific knowledge and methodologies. Indeed, it is essential to define a law that takes multiple interactions into account at different spatio-temporal scales as well as the adaptive nature of the environment, with its limitations; otherwise, the rigidity of the law is likely to generate choices or pathological behaviors in relation to ecosystems [ARN 13a].

Environmental law is a recent law that has the characteristic of being located in a continuum between public policy and regulation, with the abundant use of soft law, which provokes reactions that range from fanaticism to hostility [PHI 11]. As we have seen, it is necessary to consider the close connection between environmental law and science and implement a systematic evaluation of the analytical tools used in environmental law by recognizing the limits of our own knowledge and avoiding reductionist methods [FLO 94]. Philippopoulos-Mihalopoulos [PHI 11] called for the development of a critical environmental law that breaks with a "self-satisfied" and impoverished technicist vision and calls into question the illusion of legal control. This radical approach to environmental law is reflexive, transdisciplinary and based on a solid theoretical apparatus and methodology that needs to be developed. This device must integrate the strong connection between law and ecology.

The main difficulty of an adaptive law is to find a balance between flexibility of evolution and stability of the system, but this difficulty must not be an obstacle to the elaboration of a new generation of environmental law that remains to be invented [KIM 13, ROS 11].

Bibliography

[ABI 11] ABI-RACHED L., JOBIN M.J., KULKARNI S. *et al.*, "The shaping of modern human immune systems by multiregional admixture with archaic humans", *Science*, no. 334, pp. 89–94, 2011.

[ACH 08] ACHTMAN M., "Evolution, population structure, and phylogeography of genetically monomorphic bacterial pathogens", *Annual Review of Microbiology*, no. 62, pp. 53–70, 2008.

[ADB 11] ADB (ASIAN DEVELOPMENT BANK), Asian Judges Symposium on Environmental Decision Making, The Rule of Law, and Environmental Justice, Asian Development Bank, Philippines, Mandaluyong City, 2011.

[AGA 01] AGARWAL B., "Participatory exclusions, community forestry and gender: an analysis for South Asia and a conceptual framework", *World Development*, no. 29, pp. 1623–1648, 2001.

[AGU 95] AGUIRRE A.A., BALAZS G.H., SPRAKER T.R. *et al.*, "Adrenal and hematological responses to stress in juvenile green turtles (*Chelonia mydas*) with and without fibropapillomas", *Physiological Zoology*, no. 68, pp. 831–854, 1995.

[AKE 91] AKERELE O., HEYWOOD V., SYNGE H., *Conservation of Medicinal Plants*, Cambridge University Press, Cambridge, 1991.

[ALB 15] ALBUQUERQUE U.P., DE MEDEIROS P.M., CASAS A., *Evolutionary Ethnobiology*, Springer, London, 2015.

[ALK 02] ALKIRE S., "Dimensions of human development", *World Development*, vol. 30, pp. 181–205, 2002.

[ALL 10] ALLEN H.K., DONATO J., WANG H.H. *et al.*, "Call of the wild: antibiotic resistance genes in natural environments", *Nature Review Microbiology*, no. 8, pp. 251–259, 2010.

[ALL 17] ALLEPUZ A., DE BALOGH K., AGUANNO R. *et al.*, "Review of participatory epidemiology practices in animal health (1980–2015) and future practice directions", *PLoS ONE*, no. 12, p. e0169198, 2017.

[ALT 13] ALTIZER S., OSTFELD R.S., JOHNSON P.P. *et al.*, "Climate change and infectious diseases: from evidence to a predictive framework", *Science*, no. 341, pp. 514–519, 2013.

[ALV 14] ALVAREZ M.A., *Plant Biotechnology for Health From Secondary Metabolites to Molecular Farming*, Springer, Switzerland, 2014.

[AMI 13] AMIT D., TIWARI B.K., "Role of judiciary in environmental protection", *Journal of Environmental Resources and Development*, no. 8, pp. 200–203, 2013.

[AND 91] ANDERSON R.M., MAY R.M., *Infectious Diseases of Humans: Dynamics and Control*, Oxford University Press, Oxford, 1991.

[AND 08] ANDOW D.A., FITT G.P.P., GRAFIUS E.J. *et al.*, "Pesticide and transgenic plant resistance management in the field", in WHALON M.E., MOTA-SANCHEZ D., HOLLINGWORTH R.M. (eds), *Global Pesticide Resistance in Arthropods*, CAB International, Wallingford, 2008.

[AND 11] ANDERSON P.P.A., BERZINS I.K., FOGARTY F. *et al.*, "Sound, stress, and seahorses: the consequences of a noisy environment to animal health", *Aquaculture*, no. 311, pp. 129–138, 2011.

[ANY 12] ANYAMBA A., LINTHICUM K.J., SMALL J.L. *et al.*, "Climate teleconnections and recent patterns of human and animal disease outbreaks", *PLoS Neglected Tropical Diseases*, no. 6, p. e1465, 2012.

[APP 75] APPLETON J., *The Experience of Landscape*, John Wiley & Sons, London, 1975.

[ARA 03] ARAUJO A., JANSEN A.M., BOUCHET F. *et al.*, "Parasites, the diversity of life, and paleoparasitology", *Memorias Instituto Oswaldo Cruz*, no. 98, pp. S5–S11, 2003.

[ARI 14] ARIEY F., WITKOWSKI B., AMARATUNGA C. *et al.*, "A molecular marker of artemisinin-resistant *Plasmodium falciparum* malaria", *Nature*, no. 505, pp. 50–55, 2014.

[ARM 99] ARMSTRONG G.L., CONN L.A., PINNER R.W., "Trends in infectious disease mortality in the United States during the 20th century", *Journal of the American Medical Association*, no. 281, pp. 61–66, 1999.

[ARN 13a] ARNOLD C.A., GUNDERSON L.H., "Adaptive law and resilience", *Environmental Law Reporter*, vol. 43, pp. 10426–10443, 2013.

[ARN 13b] ARNOLD K.E., BOXALL A.B.A., BROWN A.R. *et al.*, "Assessing the exposure risk and impacts of pharmaceuticals in the environment on individuals and ecosystems", *Biology Letters*, no. 9, 2013.

[ARN 14] ARNOLD K.E., BROWN A.R., ANKLEY G.T. *et al.*, "Medicating the environment: assessing risks of pharmaceuticals to wildlife and ecosystems", *Philosophical Transactions Royal Society London B*, no. 369, p. 20130569, 2014.

[ARO 07] ARONSON J., MILTON S.J., BLIGNAUT J.N., *Restoring Natural Capital: Science, Business, and Practice*, Island Press, Washington, DC, 2007.

[ASH 98] ASHFORD R.W., CREWE W., *The Parasites of Homo sapiens: An Annotated Checklist of the Protozoa, Helminths and Arthropods For Which We Are Home*, Liverpool School of Tropical Medicine, Liverpool, 1998.

[ATL 10] ATLAS R., RUBIN C., MALOY S. *et al.*, "One Health – attaining optimal health for people, animals, and the environment", *Microbe*, no. 5, pp. 383-389, 2010.

[ATT 99] ATTFIELD R., *The Ethics of the Global Environment*, Edinburgh University Press, Edinburgh, 1999.

[AUS 07] AUSTRALIAN PUBLIC SERVICE COMMISSION, Tackling Wicked Problems: A Public Policy Perspective, Commonwealth of Australia, 2007.

[BAC 02] BACH J.-F., "Susceptibility to autoimmune and allergic diseases", *New England Journal of Medicine*, no. 347, pp. 911–920, 2002.

[BAL 82] BALLINGS J.D., FALK J.H., "Development of visual preference for natural environments", *Environment and Behavior*, no. 14, pp. 5–28, 1982.

[BAL 13] BALAZS C.L., MORELLO-FROSCH R., "The three Rs: how community-based participatory research strengthens the rigor, relevance, and reach of science", *Environmental Justice*, no. 6, pp. 9–16, 2013.

[BAR 10] BARBER J.R., CROOKS K.R., FRISTRUP K.M., "The costs of chronic noise exposure for terrestrial organisms", *Trends in Ecology and Evolution*, no. 25, pp. 180–189, 2010.

[BAR 11a] BARBIER E., *Capitalizing on Nature: Ecosystems as Natural Assets*, Cambridge University Press, Cambridge, 2011.

[BAR 11b] BARDI U., *The Limits to Growth Revisited*, Springer, New York, 2011.

[BAT 05] BATES S.L., ZHAO J.Z., ROUSH R.T. *et al.*, "Insect resistance management in GM crops: past, present and future", *Nature Biotechnology*, no. 23, pp. 57–62, 2005.

[BEA 94] BEAUCHAMP T.L., CHILDRESS J.F., *Principles of Biomedical Ethics*, Oxford University Press, Oxford, 1994.

[BEA 11] BEATLEY T., *Biophilic Cities: Integrating Nature into Urban Design and Planning*, Island Press, Washington, DC, 2011.

[BEA 14] BEAN T.G., BOXALL A.B.A., LANE J. *et al.*, "Behavioural and physiological responses of birds to environmentally relevant concentrations of an antidepressant", *Philosophical Transactions Royal Society London B*, no. 369, p. 20130575, 2014.

[BEE 16] BEERMANN J., DAMODARAN A., KIRSTEN JÖRGENSEN K. *et al.*, "Climate action in Indian cities: an emerging new research area", *Journal of Integrative Environmental Sciences*, no. 13, pp. 55–66, 2016.

[BEN 00] BENZ B.F., CEVALLOS E.J., SANTANA M.F. *et al.*, "Losing knowledge about plant use in the Sierra de Manantlan Biosphere Reserve, Mexico", *Economic Botany*, no. 54, pp. 83–191, 2000.

[BER 04] BERGSTROM C.T., LO M., LIPSITCH M., "Ecological theory suggests that antimicrobial cycling will not reduce antimicrobial resistance in hospitals", *Proceedings National Academy of Sciences USA*, no. 101, pp. 13285–13290, 2004.

[BER 08] BERKES F., *Sacred Ecology: Traditional Ecological Knowledge and Resource Management*, 2nd ed., Routledge, New York, 2008.

[BHU 12] BHULLAR K., WAGLECHNER N., PAWLOWSKI A. *et al.*, "Antibiotic resistance is prevalent in an isolated cave microbiome", *PLoS ONE*, no. 7, pp. 1–11, 2012.

[BIL 98] BILLING J., SHERMAN P.P.W., "Antimicrobial functions of spices: why some like it hot", *Quarterly Review of Biology*, no. 73, pp. 3–49, 1998.

[BIR 07] BIRD W., Natural Thinking. Investigating The Links Between The Natural Environment, Biodiversity And Mental Health, Report, The Royal Society for the Protection of Birds, London, 2007.

[BLA 00] BLANCOU J., *Histoire De La Surveillance Et Du Contrôle Des Maladies Animales Transmissibles*, OIE, Paris, 2000.

[BLO 08] BLOOM D.E., CANNING D., FINK G., "Urbanization and the wealth of nations", *Science*, no. 319, pp. 772–775, 2008.

[BOA 16] BOAKES E.H., GLIOZZO G., SEYMOUR V. *et al.*, "Patterns of contribution to citizen science biodiversity projects increase understanding of volunteers' recording behaviour", *Scientific Report*, no. 6, p. 33051, 2016.

[BOD 07a] BODEKER G., "Intellectual property rights", in BODEKER G., BURFORD G. (eds), *Traditional, Complementary and Alternative Medicine. Policy and Public Health Perspectives*, Imperial College Press, London, 2007.

[BOD 07b] BODEKER G., BURFORD G., "Medicinal plant biodiversity and local healthcare: sustainable use and livelihood development", in BODEKER G., BURFORD G. (eds), *Traditional, Complementary and Alternative Medicine. Policy and Public Health Perspectives*, Imperial College Press, London, 2007.

[BOL 11] BOLHUIOS J.J., BROWN G.R., RICHARDSON R.C. *et al.*, "Darwin in mind: new opportunities for evolutionary psychology", *PLoS Biology*, no. 9, e1001109, 2011.

[BON 08] BONNEFOY X., KAMPEN H., SWEENEY K., Public Health Significance of Urban Pests, World Health Organization, Copenhagen, 2008.

[BOR 00] BORTENSCHLAGER S., OEGGL K., *The Iceman and His Natural Environment Palaeobotanical Results*, Springer-Verlag, Wien, 2000.

[BOR 14] BORRINI-FEYERABEND G., DUDLEY N., JAEGER T. *et al.*, "Gouvernance des Aires Protégées: de la Compréhension à l'Action", Collection des lignes directrices sur les meilleures pratiques pour les aires protégées, no. 20, IUCN, Gland, 2014.

[BOU 93] BOURLIERE F., *La Conférence de la biosphère 25 ans après*, Unesco, Paris, 1993.

[BOU 16] BOUMEDIENNE S., "La colonisation du savoir", *Une Histoire Des Plantes Médicinales Du Nouveau Monde (1492–1750)*, Les Editions des Mondes à Faire, Vaulx-en-Velin, 2016.

[BOY 06] BOYLE J., "The bioethics of global biomedicine: a natural law reflection", in ENGELHARDT JR. H.T. (ed.), *Global Bioethics. The Collapse of Consensus*, M & M Scrivener P, Salem, 2006.

[BRA 10] BRADSHAW C.J.A., GIAM X., SODHI N.S., "Evaluating the relative environmental impact of countries", *PLoS ONE*, no. 5, e10440, 2010.

[BRE 13] BREARLEY G., RHODES J., BRADLEY A. *et al.*, "Wildlife disease prevalence in human-modified landscapes", *Biological Reviews of the Cambridge Philosophical Society*, no. 88, pp. 427–42, 2013.

[BRE 17] BREN D'AMOUR C., REITSMA F., BAIOCCHI G. *et al.*, "Future urban land expansion and implications for global croplands", *Proceedings of the National Academy of Sciences USA*, vol. 114, no. 34, pp. 8939–8944, 2017.

[BRO 04] BROWN T., MOON G., "From Siam to New York: Jacques May and the 'foundation' of medical geography", *Journal of Historical Geography*, vol. 30, no. 4, pp. 747–763, 2004

[BUT 06] BUTLER C.D., OLUOCH-KOSURA W., "Linking future ecosystem services and future human well-being", *Ecology and Society*, no. 11, p. 30, 2006.

[CAI 95] CAIRNS J.J., PRATT J.R., "The relationship between ecosystem health and delivery of ecosystem services", in RAPPORT D.J., GAUDET C., CALOW P.P. (eds), *Evaluating And Monitoring The Health Of Large-Scale Ecosystems*, Springer-Verlag, Heidelberg, pp. 273–294, 1995.

[CAI 14] CAI W., BORLACE S., LENGAIGNE M. *et al.*, "Increasing frequency of extreme El Niño events due to greenhouse warming", *Nature Climate Change*, no. 5, pp. 1–6, 2014.

[CAL 89] CALLICOTT J.B., AMES R.T., *Nature in Asian Traditions of Thought*, State University of New York Press, Albany, 1989.

[CAL 92] CALOW P., "Can ecosystems be healthy? Critical considerations of concepts", *Journal of Aquatic Ecosystem Health*, no. 1, pp. 1–15, 1992.

[CAL 13] CALLICOTT J.B., *Thinking Like A Planet: The Land Ethic And The Earth Ethic*, Oxford University Press, Oxford, 2013.

[CAP 98] CAPASSO L., "5300 years ago the Ice Man used natural laxatives and antibiotics", *The Lancet*, no. 352, p. 1864, 1998.

[CAR 02] CARTER R., MENDIS, K.N., "Evolutionary and historical aspects of the burden of malaria", *Clinical Microbiology Reviews*, no. 15, pp. 564–594, 2002.

[CAR 06] CARVALHO F.P.P., "Agriculture, pesticides, food security and food safety", *Environmental Science Policy*, no. 9, pp. 685–692, 2006.

[CAR 09] CARON A., GAIDET N., DE GARINE- WICHATITSKY M. *et al.*, "Evolutionary biology, community ecology and avian influenza research", *Infection Genetics and Evolution*, no. 9, pp. 298–303, 2009.

[CAR 11] CARVER S., KUENZI A., BAGAMIAN K.H. *et al.*, "A temporal dilution effect: hantavirus infection in deer mice and the intermittent presence of voles in Montana", *Oecologia*, no. 166, pp. 713–721, 2011.

[CAR 12] CARDINALE B.J., DUFFY J.E., GONZALEZ A. *et al.*, "Biodiversity loss and its impact on humanity", *Nature*, no. 486, pp. 59–67, 2012.

[CAR 14] CAROTHERS T., BRECHENMACH S., Accountability, Transparency, Participation, and Inclusion: A New Development Consensus?, Carnegie Endowment for International Peace Publications Department, 2014.

[CAS 05] CASE R.J., PAULI G.E., SOEJARTO D.D., "Factors in maintaining indigenous knowledge among ethnic communities of Manus Island", *Economic Botany*, no. 59, pp. 356–365, 2005.

[CCN 17] CCNE, Biodiversité Et Santé: Nouvelles Relations Entre l'Humanité et Le Vivant, Comité Consultatif National d'Ethique, no. 125, Avis 07/06/17, 2017.

[CEB 15] CEBALLOS G., EHRLICH A.H., EHRLICH P.R., *The Annihilation of Nature: Human Extinction of Birds and Mammals*, Johns Hopkins University Press, Baltimore, 2015.

[CHA 94] CHAMBERS R., "The origins and practice of participatory rural appraisal", *World Development*, no. 22, pp. 953–969, 1994.

[CHA 98] CHANDLER M.J., LALONDE C.E., "Cultural continuity as a hedge against suicide in Canada's First Nations", *Transcultural Psychiatry*, no. 35, pp. 193–211, 1998.

[CHA 06] CHAPMAN C.A., WASSERMAN M.D., GILLESPIE T.R. *et al.*, "Do food availability, parasitism, and stress have synergistic effects on red colobus populations living in forest fragments?", *American Journal of Physical Anthropology*, no. 131, pp. 525–534, 2006.

[CHA 09] CHANDRA A.C., *Civil Society in Search of an Alternative Regionalism in ASEAN*, International Institute for Sustainable Development, Winnipeg, 2009.

[CHA 10] CHARADEAU P., "Pour une interdisciplinarité "focalisée" dans les sciences humaines et sociales", *Questions de communication*, no. 17, pp. 195–222, 2010.

[CHA 13] CHAPMAN A., WYNDHAM J., "A human right to science", *Science*, vol. 340, no. 6138, p. 1291, 2013.

[CHA 16] CHAN K.M.A., BALVANERA P.P., BENESSAIAH K. *et al.*, "Why protect nature? Rethinking values and the environment", *Proceedings of the National Academy of Sciences USA*, no. 113, pp. 1462–1465, 2016.

[CHE 09] CHEVASSUS-AU-LOUIS B., Approche économique de la biodiversité et des services liés aux écosystèmes: Contribution à la décision publique, Centre d'Analyses Stratégiques, Paris, 2009.

[CHE 14] CHEN M., ZHANG H., LIU W. *et al.*, "The global pattern of urbanization and economic growth: evidence from the last three decades", *PLoS ONE*, no. 9, e103799, 2014.

[CHI 04] CHIVIAN E., BERNSTEIN A.S., "Embedded in nature: human health and biodiversity", *Environmental Health Perspectives*, no. 112, pp. 12–13, 2004.

[CHI 08] CHIVIAN E., BERNSTEIN A., *Sustaining Life: How Human Health Depends on Biodiversity*, Oxford University Press, Oxford, 2008.

[CHR 03a] CHRISTE P., OPPLIGER A., BANCALA F. *et al.*, "Evidence for collective medication in ants", *Ecology Letters*, no. 6, pp. 19–22, 2003.

[CHR 03b] CHRISTIN M.-S., GENDRON A.D., BROUSSEAU P.P. *et al.*, "Effects of agricultural pesticides on the immune system of *Rana pipiens* and on its resistance to parasitic infection", *Environmental Toxicological Chemistry*, no. 22, pp. 1127–1133, 2003.

[CHU 14] CHU D.K.W., POON L.L.M., GOMAA M.M. *et al.*, "Coronaviruses in Dromedary Camel herd, Saudi Arabia", *Emerging Infectious Diseases*, no. 20, pp. 1049–1054, 2014.

[CIV 15] CIVITELLO D.J., COHEN J., FATIMA H. *et al.*, "Biodiversity inhibits parasites: broad evidence for the dilution effect", *Proceedings of the National Academy of Sciences USA*, no. 112, pp. 8667–8671, 2015.

[CLA 88] CLARK L., MASON J.R., "Effect of biologically active plants used as nest material and the derived benefit to starling nestlings", *Oecologia*, no. 77, pp. 174–180, 1988.

[CLA 07] CLATWORTHY A.E., PIERSON E., HUNG D.T., "Targeting virulence: a new paradigm for antimicrobial therapy", *Nature Chemical Biology*, no. 3, pp. 541–548, 2007.

[CLE 01] CLEAVELAND S., LAURENSON M.K., TAYLOR L.H., "Diseases of humans and their domestic mammals: pathogen characteristics, host range and the risk of emergence", *Philosophical Transaction of the Royal Society of London B*, no. 356, pp. 991–999, 2001.

[CLI 14] CLINE E.H., *1177 B.C.: The Year Civilization Collapsed*, Princeton University Press, Princeton, 2014.

[COC 63] COCKBURN T.A., *The Evolution and Eradication of Infectious Diseases*, Johns Hopkins Press, Baltimore, 1963.

[COC 71] COCKBURN T. A., "Infectious disease in ancient populations", *Current Anthropology*, no. 12, pp. 45–62, 1971.

[COL 91] COLBORN T., "Epidemiology of Great Lakes bald eagles", *Journal of Toxicology and Environmental Health*, no. 33, pp. 395–453, 1991.

[COL 92] COLE S., *Making Science: Between Nature And Society*, Harvard University Press, Cambridge, 1992.

[COL 95] COLBORN T., "Environmental oestrogens: health implications for humans and wildlife", *Environmental Health Perspectives*, no. 103, pp. 135–136, 1995.

[COL 15] COLLIGNON P., ATHUKORALA P.-C., SENANAYAKE S. *et al.*, "Antimicrobial resistance: the major contribution of poor governance and corruption to this growing problem", *PLoS ONE*, no. 10, e0116746, 2015.

[COL 17] COLEMAN M., HEMINGWAY J., GLEAVE K.A. *et al.*, "Developing global maps of insecticide resistance risk to improve vector control", *Malaria Journal*, no. 16, p. 86, 2017.

[COM 99] COMBES C., MORAND S., "Do parasites live in extreme environments? Constructing hostile niches and living in them", *Parasitology*, no. 119, pp. S107–S110, 1999.

[COM 17] COMMITTEE ON THE ASSESSMENT OF THE CUMULATIVE EFFECTS OF ANTHROPOGENIC STRESSORS ON MARINE MAMMALS, Ocean Studies Board, Division on Earth and Life Studies, National Academies of Sciences, Engineering, and Medicine, 2017.

[CON 86] CONWAY G.R., Agroecosystem Analysis For Research And Development, Winrock International Institute for Agricultural Development, 1986.

[COO 95] COOPER-MARCUS C., BARNES M., *Healing Gardens: Therapeutic Benefits and Design Recommendations*, Wiley, New York, 1995.

[COO 06a] COOPER M., "Pre-empting emergence: the biological turn of the war on terror", *Theory, Culture and Society: Explorations in Critical Social Science*, no. 23, pp. 113–135, 2006.

[COO 06b] COOPER M.D., ALDER M.N., "The evolution of adaptive immune systems", *Cell*, no. 124, pp. 815–822, 2006.

[COR 09] CORRAL-VERDUGO V., BONNES M., TAPIA-FONLLEM C. *et al.*, "Correlates of pro-sustainability orientation: the affinity towards diversity", *Journal of Environmental Psychology*, no. 29, pp. 34–43, 2009.

[COR 13] CORBEL V., N'GUESSAN R., "Distribution, mechanisms, impact and management of insecticide resistance in malaria vectors: a pragmatic review", in MANGUIN S. (ed.), *Anopheles Mosquitoes: New Insights into Malaria Vectors*, InTech, Rijeka, 2013.

[COS 92] COSTANZA R., NORTON B.G., HASKELL B.D., *Ecosystem Health: New Goals for Environmental Management*, Island Press, Washington, 1992.

[COS 97a] COSTANZA R., CUMBERLAND J., DALY H. *et al.*, *An Introduction To Ecological Economics*, CRC Press, Boca Raton, 1997.

[COS 97b] COSTANZA R., D'ARGE R., DE GROOT R. *et al.*, "The value of the world's ecosystem services and natural capital", *Nature*, no. 387, pp. 253–260, 1997.

[COS 99a] COSTANZA R., MAGEAU M., "What is a healthy ecosystem?", *Aquatic Ecology*, no. 33, pp. 105–115, 1999.

[COS 07] COSTANZA R., GRAUMLICH L.J., STEFFEN W., *Sustainbility or Collapse? An Integrated History and Future of People on Earth*, Massachusetts Institute of Technology, Cambridge, 2007.

[COS 14] COSTANZA R., DE GROOT R., PAUL SUTTON P.P. *et al.*, "Changes in the global value of ecosystem services", *Global Environmental Change*, no. 26, pp. 152–158, 2014.

[COU 07] COURNIL C., MAZZEGA P., "Réflexions prospectives sur une protection juridique des réfugiés écologiques", *Revue européenne des migrations internationales*, no. 23, p. 1, 2007.

[COU 17] COURTIER-ORGOGOZO V., MORIZOT B., BOËTE C., "Agricultural pest control with CRISPR-based gene drive. Time for public debate: should we use gene drive for pest control?", *EMBO Reports*, no. 18, pp. 878–880, 2017.

[CRA 08] CRABBÉ P.J., MANNO J.P., "Ecological integrity as an emergent global public good", in WENSTRA L., BOSSELMANN K., WESTRA R. (eds), *Reconciling Human Existence with Ecological Integrity: Science, Ethics, Economics and Law*, Earthscan, London, 2008.

[CRE 06] CREWS D., MCLACHLAN J.A., "Epigenetics, evolution, endocrine disruption, health, and disease", *Endocrinology*, no. 147, pp. 4–10, 2006.

[CRO 03] CROSBY A.W., *The Columbian Exchange: Biological and Cultural Consequences of 1492*, Wesport, Praeger, 2003.

[CUI 15] CUI L., MHARAKURWA S., NDIAYE D. *et al.*, "Antimalarial drug resistance: literature review and activities and findings of the ICEMR network", *American Journal of Tropical Medicine and Hygiene*, no. 93, pp. 57–68, 2015.

[CUN 93] CUNNINGHAM A.B., African medicinal plants: setting priorities at the interface between conservation and primary health care n people and plants initiative, Working paper I, UNESCO, Nairobi, 1993.

[CUN 01] CUNNINGHAM A.B., *Applied Ethnobotany: People, Wild Plant Use And Conservation*, Routledge Earthscan, London, 2001.

[CUR 01] CURTIS V., BIRAN A., "Dirt, disgust, and disease: is hygiene in our genes?", *Perspectives in Biology and Medicine*, no. 44, pp. 17–31, 2001.

[CUT 11] CUTHBERT R.J., TAGGART M.A., PRAKASH V. *et al.*, "Effectiveness of action in India to reduce exposure of gyps vultures to the toxic veterinary drug diclofenac", *PLoS ONE*, no. 6, e19069, 2011.

[CUT 14] CUTHBERT R.J., TAGGART M.A., PRAKASH V. *et al.*, "Avian scavengers and the threat from veterinary pharmaceuticals", *Philosophical Transactions of the Royal Society London B*, no. 369, 20130574, 2014.

[DAD 15] DADVAND P., NIEUWENHUIJSEN M.J., ESNAOLA M. *et al.*, "Green spaces and cognitive development in primary schoolchildren", *Proceedings of the National Academy of Sciences USA*, no. 112, pp. 7937–7942, 2015.

[DAG 14] DAGNE T., "Protecting traditional knowledge in international intellectual property law: imperatives for protection and choice of modalities", *J. Marshall Review of Intellectual Property Law*, vol. 25, pp. 25–49, 2014.

[DAI 96] DAILY G.C., EHRLICH P.P.R., "Global change and human susceptibility to disease", *Annual Review of Energy and Environment*, no. 21, pp. 125–144, 1996.

[DAI 97] DAILY G.C., *Nature's Services: Societal Dependence on Natural Ecosystems*, Island Press, Washington, DC, 1997.

[DAN 08] DANILOVA N., "Evolution of the human immune system", *Encyclopedia of Life Sciences*, John Wiley & Sons, Chichester, 2008.

[DAN 15] D'ANNUNZIO R., SANDKER M., FINEGOLD Y. *et al.*, "Projecting global forest area towards 2030", *Forest Ecology and Managenent*, no. 352, pp. 124–133, 2015.

[DAS 09] DAS D., PHYO A.P., TARNING J. *et al.*, "Artemisinin resistance in *Plasmodium falciparum* malaria", *New England Journal of Medicine*, no. 361, pp. 455–467, 2009.

[DAV 08] DAVIES T.J., PEDERSEN A.B., "Phylogeny and geography predict pathogen community similarity in wild primates and humans", *Proceedings of the Royal Society of London B*, no. 275, pp. 1695–1701, 2008.

[DAV 15] DADVAND P., NIEUWENHUIJSEN M.J., ESNAOLA M., *et al.*, "Green spaces and cognitive development in primary schoolchildren", *Proceedings of the National Academy of Sciences*, USA, no. 112, pp. 7937–7942, 2015.

[DCO 11] D'COSTA V.M., KING C.E., KALAN L. *et al.*, "Antibiotic resistance is ancient", *Nature*, no. 477, pp. 457–461, 2011.

[DEB 93] DE BOER R.J., PERELSON A.S., "How diverse should the immune system be?" *Proceedings Biological Society*, no. 252, pp. 171–175, 1993.

[DEG 02] DE GROOT R.S., WILSON M.A., BOUMANS R.M.J., "A typology for the classification, description and evaluation of ecosystem functions, goods and services", *Ecological Economics*, no. 41, pp. 393–408, 2002.

[DEL 97] DE LEO G.A., LEVIN S., "The multifaceted aspects of ecosystem integrity", *Conservation Ecology*, no. 1, pp. 1–3, 1997.

[DEL 00] DE LAINE M., *Fieldwork, Participation and Practice: Ethics and Dilemmas in Qualitative Research*, SAGE Publications, London, 2000.

[DEL 08] DE LA ROCQUE S., MORAND S., HENDRIX G. (eds), "Climate change and pathogens", *Revue Scientifique et Technique*, vol. 27, no. 2, 2008.

[DEN 95] DENNETT D.C., *Darwin's Dangerous Idea: Evolution and the Meanings of Life*, Simon & Schuster, London, 1995.

[DER 11] DERNE B.T., FEARNLEY E.J., LAU C.L. *et al.*, "Biodiversity and leptospirosis risk: a case of pathogen regulation?", *Medical Hypotheses*, no. 77, pp. 339–344, 2011.

[DER 13] DE ROODE J.C., LEFÈVRE T., HUNTER M.D., "Self-medication in animals", *Science*, no. 340, pp. 5–6, 2013.

[DES 05] DESCOLA P., *Par-delà Nature et Culture*, Gallimard, Paris, 2005.

[DES 11] DESCOLA P., *L'Ecologie des Autres. L'Anthropologie et la Question de la Nature*, Quae, Versailles, 2011.

[DIA 05] DIAMOND J.M., *Collapse: How Societies Choose To Fail Or Succeed*, Penguin Group, New York, 2005.

[DIA 15] DIAZ S., DEMISSEW S., CARABIAS J. *et al.*, "The IPBES conceptual framework – connecting nature and people", *Current Opinion in Environmental Sustainability*, no. 14, pp. 1–16, 2015.

[DÍA 15] DÍAZ S., DEMISSEW S., JOLY C. *et al.*, "A Rosetta Stone for Nature's benefits to people", *PLoS Biology*, no. 13, e1002040., 2015.

[DIC 06] DICKE M., TAKKEN W., *Chemical Ecology: From Gene to Ecosystem*, Springer, New York, 2006.

[DIE 09] DIENER E., *Well-being for Public Policy*, Oxford University Press, Oxford, 2009.

[DIL 08] DILLARD J., "A slaughterhouse nightmare: psychological harm suffered by slaughterhouse employees and the possibility of redress through legal reform", *Georgetown Journal on Poverty Law & Policy*, no. 15, pp. 391–408, 2008.

[DIR 91] DIRECTEUR GENERAL DE L'OMS, Un paradigme pour la santé: un cadre pour une nouvelle action publique, EB89/11, 9 December 1991.

[DOB 95] DOBSON A.P., "The ecology and epidemiology of rinderpest virus in Serengeti and Ngorongoro crater conservation area", in SINCLAIR A.R.E., ARCESE P.P. (eds), *Serengetti II. Research Management and Conservation of an Ecosystem*, University of Chicago Press, Chicago, 1995.

[DOD 14] DODSON B.L., GRANT H., OLUWATOBI P.P., "*Wolbachia* enhances West Nile Virus (WNV) infection in the mosquito *Culex tarsalis*", *PLoS ONE*, no. 8, e2965, 2014.

[DOM 11] DOMINGUEZ-BELLO M.G., BLASER M.J., "The human microbiota as a marker for migrations of individuals and populations", *Annual Reviews of Anthropology*, no. 40, pp. 451–474, 2011.

[DON 11] DONALDSON S., KYMLICKA W., *Zoopolis. A Political Theory of Animal Rights*, Oxford University Press, Oxford, 2011.

[DON 13] DONOVAN G.H., BUTRY D.T., MICHAEL Y.L. *et al.*, "The relationship between trees and human health: Evidence from the spread of the Emerald ash borer", *American Journal of Preventive Medicine*, no. 44, pp. 139–145, 2013.

[DOW 98] DOWER N., *World Ethics – The New Agenda*, Edinburgh University Press, Edinburgh, 1998.

[DRI 09] DRISCOLL C., MACDONALD D.W., O'BRIEN S.J., "From wild animals to domestic pets, an evolutionary view of domestication", *Proceedings of the National Academy of Sciences USA*, no. 106, pp. 9971–9978, 2009.

[DUA 12] DUARTE TAGLES H., IDROVO A.J., "Biodiversity and mental health", in LAMEED G.A. (ed.), *Biodiversity Enrichment in a Diverse World*, InTech, Rijeka, 2012.

[DUN 09] DUNN R.R., HARRIS N.C., COLWELL R.K. *et al.*, "The sixth mass coextinction: are most endangered species parasites and mutualists?", *Proceedings of the Royal Society of London B: Biological Sciences*, no. 276, pp. 3037–3045, 2009.

[DUN 10a] DUNN R.R., "Global mapping of ecosystem disservices: the unspoken reality that nature sometimes kills us", *Biotropica*, no. 42, pp. 555–557, 2010.

[DUN 10b] DUNN R.R., DAVIES T.J., HARRIS N.C. *et al.*, "Global drivers of human pathogen richness and prevalence", *Proceedings of the Royal Society of London B: Biological Sciences*, no. 277, pp. 2587–2595, 2010.

[DUP 04] DUPUY J.P., *Pour un Catastrophisme Éclairé: Quand l'Impossible est Certain*, Le Seuil, Paris, 2004.

[DUR 06] DURON O., LABBÉ P., BERTICAT C. *et al.*, "High *Wolbachia* density correlates with cost of infection for insecticide resistant *Culex pipiens mosquitoes*", *Evolution*, no. 60, pp. 303–314, 2006.

[DUT 98] DUTTA P., KHAN S.A., SHARMA C.K. *et al.*, "Distribution of potential dengue vectors in major townships along the national highways and trunk roads of northeast India", *Southeast Asian Journal Tropical Medicine and Public Health*, no. 29, pp. 173–176, 1998.

[DUT 00] DUTFIELD G., *Intellectual Property Rights, Trade and Biodiversity: Seeds and Plant Varieties*, Routledge Earthscan, London, 2000.

[DYE 14] DYE C., "After 2015: infectious diseases in a new era of health and development", *Philosophical Transactions of the Royal Society London B: Biological Sciences*, no. 369, 20130426, 2014.

[EAS 01] EASTERLIN R.A., "Income and happiness: towards a unified theory", *Economic Journal*, no. 111, pp. 465–484, 2001.

[EHR 68] EHRLICH P.R., *The Population Bomb*, Sierra Club, Ballantine Books, San Francisco, 1968.

[EHR 12] EHRLICH P.R., KAREIVA P.M., DAILY G.C., "Securing natural capital and expanding equity to rescale civilization", *Nature*, no. 486, pp. 68–73, 2012.

[EIS 95] EISNER T., MEINWALD J., *Chemical Ecology: The Chemistry of Biotic Interaction*, National Academy Press, Washington D.C., 1995.

[ELM 13] ELMQVIST T., FRAGKIAS M., GOODNESS J. *et al.*, *Urbanization, Biodiversity and Ecosystem Services: Challenges and Opportunities A Global Assessment*, Springer, London, 2013.

[ELT 58] ELTON C.S., *The Ecology of Invasions by Animals and Plants*, Methuen, London, 1958.

[ENG 02] ENGEL C., *Wild Health*, Weidenfeld and Nicholson, London, 2002.

[ENG 06a] ENGELHARDT JR. H.T., "Global bioethics: an introduction to the collapse of consensus", in ENGELHARDT JR. H.T. (ed.), *Global Bioethics. The Collapse of Consensus*, M & M Scrivener P, Salem, 2006.

[ENG 06b] ENGELHARDT JR. H.T., "The search for a global morality: bioethics, the culture wars, and moral diversity", in ENGELHARDT JR. H.T., (ed.), *Global Bioethics. The Collapse of Consensus*", M & M Scrivener P, Salem, 2006.

[ESC 11] ESCOBEDO F.J., KROEGER T., WAGNER J.E., "Urban forests and pollution mitigation: analysing ecosystem services and disservices", *Environmental Pollution*, no. 159, pp. 2078–2087, 2011.

[ETK 11] ETKIN N.L., TICKTIN T., MCMILLEN H.L., "Ethnoecological approaches to integrating theory and method in ethnomedical research", in ANDERSON E.N., PEARSALL D., HUNN E. *et al.* (eds), *Ethnobiology*, John Wiley & Sons, London, 2011.

[EVA 15] EVANS H., MITRE E., "Worms as therapeutic agents for allergy and asthma: understanding why benefits in animal studies have not translated into clinical success", *Journal of Allergy and Clinical Immunology*, no. 135, pp. 343–353, 2015.

[EWE 13] EWERS R.M., DIDHAM R.K., PEARSE W.D. *et al.*, "Using landscape history to predict biodiversity patterns in fragmented landscapes", *Ecology Letters*, no. 16, pp. 1221–1233, 2013.

[FAD 86] FADEN R.R., *A History and Theory of Informed Consent*, Oxford University Press, 1986.

[FAG 09] FAGAN B., *Floods, Famines, and Emperors*, Basic Books, New York, 2009.

[FAL 03] FALUSH D., WIRTH T., LINZ B. *et al.*, "Traces of human migrations in *Helicobacter pylori* populations", *Science*, no. 299, pp. 1582–1585, 2003.

[FAO 06] FAO, Liverstock's Long Shadow: Environmental Issues and Options, Food and Agriculture Organization of the United Nations, Rome, 2006.

[FAO 10] FAO, OIE, OMS, Collaboration FAO–OIE–OMS, Note tripartite conceptuelle: Partage des responsabilités et coordination des actions globales pour gérer les risques sanitaires à l'interface animal-homme-écosystèmes, 2010.

[FAO 12] FAO, World agriculture towards 2030/2050 – The 2012 revision, ESA Working paper No. 12–03, Food and Agriculture Organization of the United Nations, Rome, 2012.

[FAO 15] FAO, The Second Report on the State of the World's Animal Genetics Resources for Food and Agriculture, Food and Agriculture Organization of the United Nations, Rome, 2015.

[FER 16] FERNANDES M.R., MOURA Q., SARTORI L. *et al.*, "Silent dissemination of colistin-resistant Escherichia coli in South America could contribute to the global spread of the mcr-1 gene", *Euro Surveillance*, vol. 21, no. 17, 2016.

[FIG 14] FIGGE L., MARTENS P., "Globalisation continues: the Maastricht Globalisation Index revisited and updated", *Globalizations*, vol. 11, no. 6, pp. 875–893, 2014.

[FIN 08] FINCHER C., THORNHILL R., "A parasite-driven wedge: infectious diseases may explain language and other biodiversity", *Oikos*, no. 9, pp. 1289–1297, 2008.

[FIS 00] FISCHER F., *Citizens, Experts, and The Environment The Politics of Local Knowledge*, Duke University Press, Durham, 2000.

[FIS 09] FISHER M.C., GARNER T.W.J., WALKER S.F., "Global emergence of *Batrachochytrium dendrobatidis* and amphibian Chytridiomycosis in space, time, and host", *Annual Review of Microbiology*, no. 63, pp. 291–310, 2009.

[FIS 12] FISHER M.C., HENK D.A., BRIGGS C.J. *et al.*, "Emerging fungal threats to animal, plant and ecosystem health", *Nature*, no. 484, pp. 186–194, 2012.

[FIS 16] FISCHER-KOWALSKI M., HELGA WEISZ H., "The archipelago of social ecology and the island of the Vienna School", in HABERL H., FISCHER-KOWALSKI M., KRAUSMANN F. (eds), *Social Ecology Society-Nature Relations across Time and Space*, Springer, Switzerland, pp. 3–28, 2016.

[FLA 09] FLAJNIK M.F., KASAHARA M., "Origin and evolution of the adaptive immune system: genetic events and selective pressures", *Nature Review Genetics*, no. 11, pp. 47–59, 2009.

[FLE 06] FLEMING, J.O., COOK, T.D., "Multiple sclerosis and the hygiene hypothesis", *Neurology*, no. 67, pp. 2085–2086, 2006.

[FLO 94] FLOURNOY A., "Coping with complexity", *Loyola of Los Angeles Law Review*, vol. 27, pp. 809–824, 1994.

[FOR 12] FORSBERG K.J., REYES A., WANG B. *et al.*, "The shared antibiotic resistome of soil bacteria and human pathogens", *Science*, no. 337, pp. 1107–1111, 2012.

[FOS 97] FOSTER J., *Valuing Nature? Ethics, Economics and the Environment*, Routledge, Oxon, 1997.

[FOS 10] FOSTER J.B., CLARK B., YORK R., *The Ecological Rift: Capitalism's War On The Earth*, Monthly Review Press, New York, 2010.

[FRA 59] FRAENKEL G.S., "Raison d'être of secondary plant substances", *Science*, no. 129, pp. 1466–1470, 1959.

[FRA 96] FRANK S.A., "Models of parasite virulence", *Quarterly Review of Biology*, no. 71, pp. 37–78, 1996.

[FRA 12] FRANCIS C.D., KLEIST N.J., ORTEGA C.P.P. *et al.*, "Noise pollution alters ecological services: enhanced pollination and disrupted seed dispersal", *Proceedings of the Royal Society of London B: Biological Sciences*, no. 279, pp. 2727–2735, 2012.

[FRI 06] FRIEND M., Disease Emergence and Resurgence: The Wildlife-Human Connection, Circular 1285, U.S. Geological Survey, Reston, 2006.

[FUN 93] FUNTOWICZ S., RAVETZ J., "Science for the post-normal age", *Futures*, no. 25, pp. 739–755, 1993.

[FUR 10] FURUSE Y., SUZUKI, A., OSHITANI H., "Origin of measles virus: divergence from rinderpest virus between the 11th and 12th centuries", *Virology Journal*, no. 7, p. 52, 2010.

[GAG 07] GAGNEUX S., SMALL, P.M., "Global phylogeography of *Mycobacterium tuberculosis* and implications for tuberculosis product development", *Lancet Infectious Diseases*, no. 7, pp. 328–337, 2007.

[GAL 09] GALE P., DREW T., PHIPPS L.P. *et al.*, "The effect of climate change on the occurrence and prevalence of livestock diseases in Great Britain: a review", *Journal of Applied Microbiology*, no. 106, pp. 1409–1423, 2009.

[GAR 01] GARRETT, L., *Betrayal of Trust: The Collapse of Global Health*, Oxford University Press, Oxford, 2001.

[GAU 14] GAULT G., "Quels questionnements sanitaires en milieu urbain?", in MORAND S., MOUTOU F., RICHOMME C. (eds), *Faune Sauvage, Biodiversité et Santé, Quels Défis?*, Quae, Versailles, 2014.

[GEI 98] GEISSLER P.W., "Worms are our life', part I: understandings of worms and the body among the Luo of western Kenya", *Anthropology & Medicine*, no. 5, pp. 63–79, 1998.

[GEL 16] GELLER A.M., *Environmental Justice Research Roadmap*, United States Environmental Protection Agency, Washington, D.C., 2016.

[GES 03] GESLER W., *Healing Places*, Rowman & Littlefield, Baltimore, 2003.

[GET 10] GETHING P.W., SMITH D.L., PATIL A.P. *et al.*, "Climate change and the global malaria recession", *Nature*, no. 465, pp. 342–344, 2010.

[GIB 10] GIBBS H.K., RUESCH A.S., ACHARD F. *et al.*, "Tropical forests were the primary sources of new agricultural land in the 1980s and 1990s", *Proceedings of the National Academy of Sciences USA*, no. 107, pp. 16732–16737, 2010.

[GIL 94] GILLON R., "Medical ethics: four principles plus attention to scope", *British Medical Journal*, no. 309, pp. 184–188, 1994.

[GIL 16] GILL G., "Environmental Justice in India: The National Green Tribunal and Expert Members", *Transnational Environmental Law*, no. 5, pp. 175–205, 2016.

[GLA 05] GLASER R., KIECOLT-GLASER J.K., "Science and society: stress-induced immune dysfunction: implications for health", *Nature Reviews Immunology*, no. 5, pp. 243–251, 2005.

[GLO 96] GLOWKA L., BURHENNE-GUILMIN F., SYNGE H., Guide de la Convention sur la diversité biologique, Environmental Policy and Law Paper no. 30, UICN, Gland & Cambridge, 1996.

[GLO 15] GLOTFELTY C., QUESNEL E., *The Biosphere and the Bioregion: Essential Writings of Peter Berg*, Routledge, Oxon, 2015.

[GOL 15] GOLDBERG T.L., PATZ J.A., "The need for a global health ethic", *The Lancet*, no. 386, pp. 38–39, 2015.

[GÓM 13a] GÓMEZ J.M., NUNN C.L., VERDÚ M., "Centrality in primate-parasite networks reveals the potential for the transmission of emerging infectious diseases to humans", *Proceedings of the National Academy of Science of the USA*, no. 110, pp. 7738–7741, 2013.

[GÓM 13b] GÓMEZ A., NICHOLS E., "Neglected wild life: parasitic biodiversity as a conservation target", *International Journal of Parasitology: Parasites Wildlife*, no. 2, pp. 222–227, 2013.

[GOT 14] GOTTDENKER N.L., STREICKER D.G., FAUST C.L. *et al.*, "Anthropogenic land use change and infectious diseases: a review of the evidence", *EcoHealth*, vol. 11, pp. 619–632, 2014.

[GOÜ 08] GOÜY DE BELLOCQ J., CHARBONNEL N., MORAND S., "Coevolutionary relationship between helminth diversity and MHC class II polymorphism in rodents", *Journal of Evolutionary Biology*, no. 21, pp. 1144–1150, 2008.

[GRA 09] GRADÉ J.T., TABUTI J.R.S., VAN DAMME P.P., "Ethnoveterinary knowledge in pastoral Karamoja, Uganda", *Journal of Ethnopharmacology*, no. 122, pp. 273–293, 2009.

[GRA 12] GRACE D., MUTUA F., OCHUNGO P.P. *et al.*, Mapping of poverty and likely zoonoses hotspots, Zoonoses Project 4, Report to the UK Department for International Development, ILRI, Nairobi, 2012.

[GRA 15] GRANATO A., CHEN Y., WESEMANN D.R., "Primary immunoglobulin repertoire development: time and space matter", *Current Opinion in Immunology*, no. 33, pp. 126–131, 2015.

[GUA 10] GUAL M.A., NORGAARD R.B., "Bridging ecological and social systems coevolution: a review and proposal", *Ecological Economics*, no. 69, pp. 707–717, 2010.

[GUE 04] GUERNIER V., HOCHBERG M.E., GUÉGAN J.-F., "Ecology drives the worldwide distribution of human diseases", *PLoS Biology*, no. 2, pp. 740–746, 2004.

[GUI 11] GUIS H., CAMINADE C., CALVETE C. *et al.,* "Integration of climate and disease models gives mechanistic insight into vector-borne disease emergence", *Journal Royal Society Interface*, no. 9, pp. 339–350, 2011.

[GUL 00] GULLONE E., "The Biophilia hypothesis and life in the 21st century: increasing mental health or increasing pathology?", *Journal of Happiness Studies*, no. 1, pp. 293–321, 2000.

[GUN 09] GUNDERSON L.H., HOLLING C.S., *Panarchy: Understanding Transformations in Human and Natural Systems*, Island Press, Washington, DC, 2009.

[GÜN 17] GÜNERALP B., ZHOU Y., ÜRGE-VORSATZ D. *et al.*, "Global scenarios of urban density and its impacts on building energy use through 2050", *Proceedings of the National Academy of Sciences USA*, doi:10.1073/pnas. 1606035114, 2017.

[GUT 05] GUTIERREZ M.C., BRISSE S., BROSCH R. *et al.*, "Ancient origin and gene mosaicism of the progenitor of Mycobacterium tuberculosis", *PLoS Pathogens*, no. 1, e5, 2005.

[HAA 13] HAAHTELA T., HOLGATE S., PAWANKAR R. *et al.*, "The biodiversity hypothesis and allergic disease: world allergy organization position statement". *World Allergy Organization Journal*, vol. 6, no. 3, doi:10.1186/1939-4551-6-3, 2013.

[HAB 07] HABERL H., ERB K.-H., KRAUSMANN F. *et al.*, "Quantifying and mapping the human appropriation of net primary production in earth's terrestrial ecosystems", *Proceedings of the National Academy of Sciences USA*, no. 104, pp. 12942–12947, 2007.

[HAD 15] HADDAD N.M., BRUDVIG L.A., CLOBERT J. *et al.*, "Habitat fragmentation and its lasting impact on earth's ecosystems", *Science Advances*, no. 1, e1500052, 2015.

[HAL 05] HALBERSTEIN R.A., "Medicinal plants: historical and cross-cultural usage patterns", *Annals of Epidemiology*, no. 15, pp. 686–699, 2005.

[HAL 07] HALPIN K., HYATT A.D., PLOWRIGHT R.K. *et al.,* "Emerging viruses: coming in on a wrinkled wing and a prayer", *Clinical Infectious Disease*, no. 44, pp. 711–717, 2007.

[HAM 16a] HAMMER T.J., FIERER N., HARDWICK B. *et al.*, "Treating cattle with antibiotics affects greenhouse gas emissions, and microbiota in dung and dung beetles", *Proceedings of the Royal Society of London B: Biological Sciences*, no. 283, 20160150, http://dx.doi.org/10.1098/rspb.2016.0150, 2016.

[HAM 16b] HAMMOND A., GALIZI R., KYROU K. *et al.*, "A CRISPR-Cas9 gene drive system targeting female reproduction in the malaria mosquito vector *Anopheles gambiae*", *Nature Biotechnology*, no. 34, pp. 78–83, 2016.

[HAN 12] HANSKI I., VON HERTZEN L., FYHRQUIST N. *et al.*, "Environmental biodiversity, human microbiota, and allergy are interrelated", *Proceedings of the National Academy of Sciences USA*, no. 109, pp. 8334–8339, 2012.

[HAR 05] HART B.L., "The evolution of herbal medicine: behavioural perspectives", *Animal Behaviour*, no. 70, pp. 975–989, 2005.

[HAR 11] HART B.L., "Behavioural defences in animals against pathogens and parasites: parallels with the pillars of medicine in humans", *Philosophical Transactions of the Royal Society of London B: Biological Sciences*, no. 366, pp. 3406–3417, 2011.

[HAS 17] HASSELL J.M., BEGON M., WARD M.J. *et al.*, "Urbanization and disease emergence: dynamics at the wildlife–livestock–human interface", *Trends in Ecology and Evolution*, no. 32, pp. 55–67, 2017.

[HAT 12] HATCHER M.J., DICK J.T.A., DUNN A., "Disease emergence and invasions", *Functional Ecology*, no. 26, pp. 1275–1287, 2012.

[HAY 10] HAYES T.B., FALSO P., GALLIPEAU S. *et al.*, "The cause of global amphibian declines: a developmental endocrinologist's perspective", *Journal of Experimental Zoology*, no. 213, pp. 921–933, 2010.

[HEM 00] HEMINGWAY J., RANSON H., "Insecticide resistance in insect vectors of human disease", *Annual Review of Entomology*, no. 45, pp. 371–391, 2000.

[HEN 12] HENN B.M., CAVALLI-SFORZA L.L., FELDMAN M.W., "The great human expansion", *Proceedings of the National Academy of Sciences USA*, no. 109, pp. 17758–17764, 2012.

[HEY 06] HEYMANN D.L., "SARS and emerging infectious diseases: a challenge to place global solidarity above national sovereignty", *Annals of the Academy of Medicine*, no. 35, pp. 350–353, 2006.

[HEY 14] HEYLEN M., RUYSSERS, N.E., GIELIS, E.M. *et al.*, "Of worms, mice and man: an overview of experimental and clinical helminth-based therapy for inflammatory bowel disease", *Pharmacology and Therapeutics*, no. 143, pp. 153–167, 2014.

[HIN 16] HING S., NARAYAN E.J., THOMPSON R.C.A. *et al.*, "The relationship between physiological stress and wildlife disease: consequences for health and conservation", *Wildlife Research*, no. 41, pp. 51–60, 2016.

[HIR 13] HIRABAYASHI Y., MAHENDRAN R., KOIRALA S. *et al.*, "Global flood risk under climate change", *Nature Climate Change*, no. 3, pp. 816–821, 2013.

[HOB 01] HOBERG E.P., ALKIRE N.L., DE QUEIROZ A. *et al.*, "Out of Africa: origins of the *Taenia* tapeworms in humans", *Proceedings of the Royal Society of London B: Biological Sciences*, no. 268, pp. 781–787, 2001.

[HOF 11] HOFFMANN A.A., MONTGOMERY B.L., POPOVICI J. *et al.*, "Successful establishment of *Wolbachia* in *Aedes populations* to suppress dengue transmission", *Nature*, no. 476, pp. 454–457, 2011.

[HOL 86] HOLLING C.S., "The resilience of terrestrial ecosystems: local surprise and global change", in CLARK W.C., MUNN R.E. (eds), *Sustainable Development of the Biosphere*, Cambridge University Press, Cambridge, 1986.

[HOL 16] HOLMES A.H., MOORE L.S.P., SUNDSFJORD A. *et al.*, "Understanding the mechanisms and drivers of antimicrobial resistance", *Lancet*, no. 387, pp. 176–187, 2016.

[HOO 92] HOOKER C.A., "Responsibility, ethics and nature", in COOPER D.E., PALMER J.A. (eds), *The Environment in Question: Ethics and Global Issues*, Routledge, London, 1992.

[HOS 10] HOSTE H., SOTIRAKI S., LANDAU S.Y. *et al.*, "Goat-nematode interactions: think differently", *Trends in Parasitology*, no. 26, pp. 376–381, 2010.

[HOW 11] HOWELL A.J., DOPKO R.L., PASSMORE H.-A. *et al.*, "Nature connectedness: associations with well-being and mindfulness", *Personality and Individual Differences*, no. 51, pp. 166–171, 2011.

[HSU 06] HSU E., "Reflections on the "discovery" of the antimalarial qinghao", *British Journal of Clinical Pharmacology*, no. 61, pp. 666–670, 2006.

[HUB 11] HUBER M., KNOTTNERUS J.A., GREEN L. *et al.*, "How should we define health?", *British Medical Journal*, no. 343, d4163, 2011.

[HUD 17] HUDSON L.N., TIM NEWBOLD T., CONTU S. *et al.*, "The database of the PREDICTS (Projecting Responses of Ecological Diversity In Chang in Terrestrial Systems) project", *Ecology and Evolution*, no. 7, pp. 145–188, 2017.

[HUF 96a] HUFFMAN M.A., KOSHIMIZU K., OHIGASHI H., "Ethnobotany and zoopharmacognosy of Vernonia amygdalina, a medicinal plant used by humans and chimpanzees", in CALIGARI P.D.S., HIND D.J.N. (eds), *Compositae: Biology and Utilization*, vol. 2, Royal Botanical Gardens, Kew, 1996.

[HUF 96b] HUFFMAN M.A., PAGE J.E., SUKHDEO M.V.K. *et al.*, "Leaf-swallowing by chimpanzees, a behavioral adaptation for the control of strongyle nematode infections", *International Journal of Primatology*, no. 17, pp. 475–503, 1996.

[HUF 97] HUFFMAN M.A., "Current evidence of self-medication in primates: a multidisciplinary perspective", *Yearbook of Physical Anthropology*, no. 40, pp. 171–200, 1997.

[HUF 01a] HUFFMAN M.A., "Self-medicative behavior in the African Great Apes: an evolutionary perspective into the origins of human traditional medicine", *Bioscience*, no. 51, pp. 651–661, 2001.

[HUF 01b] HUFFMAN M.A., CATON J.M., "Self-induced increase of gut motility and the control of parasite infections in wild chimpanzees", *International Journal of Primatology*, no. 22, pp. 329–346, 2001.

[HUT 03] HUTCHINGS M.R., ATHANASIADOU S., KYRIAZAKIS I. *et al.*, "Can animals use foraging behaviour to combat parasites?", *Proceedings of the Nutrition Society*, no. 62, pp. 361–370, 2003.

[IBI 16] IBISCH P.L., HOFFMANN M.T., KREFT S. *et al.*, "A global map of roadless areas and their conservation status", *Science*, no. 354, pp. 1423–1427, 2016.

[IMH 06] IMHOFF M.L., BOUNOUA L., "Exploring global patterns of net primary production carbon supply and demand using satellite observations and statistical data", *Journal of Geophysical Research*, no. 111, D22S12, 2006.

[IPB 16] IPBES (INTERGOVERNMENTAl SCIENCE-POLICY PLATFORM ON BIODIVERSITY AND ECOSYSTEM SERVICES), "The methodological assessment report on scenarios and models of biodiversity and ecosystem services", in FERRIER S., NINAN K.N., LEADLEY P. *et al.* (eds), *Secretariat of the Intergovernmental Science-Policy Platform on Biodiversity and Ecosystem Services*, available at: https://www.ipbes.net/sites/ default/files/downloads/ pdf/2016.methodological_assessment_report_scenarios_models.pdf, Bonn, 2016.

[ISB 06] ISBELL L.A., "Snakes as agents of evolutionary change in primate brain", *Journal of Human Evolution*, no. 51, pp. 1–35, 2006.

[IUC 17] IUCN, World declaration on the environmental rule of law, adopted at the IUCN 1st World Congress on Environmental Law, IUCN World Commission on Environmental Law & the United Nations Environment Programme, April 2016 in Rio de Janeiro, Brazil, Finalised in February 2017.

[JAM 16] JAMES P., HART J.E., BANAY R.F. *et al.*, "Exposure to greenness and mortality in a nationwide prospective cohort study of women", *Environmental Health Perspectives*, no. 124, pp. 1344–1352, 2016.

[JAN 78] JANZEN D.H., "Complications in interpreting the chemical defense of trees against tropical arboreal plant-eating vertebrates", in MONTGOMERY G.G. (ed.), *The Ecology of Arboreal Folivores*, Smithsonian Institution Press, Washington, DC, pp. 73–84, 1978.

[JAY 15] JAYME S.I., FIELD H.E., DE JONG C. *et al.*, "Molecular evidence of Ebola Reston virus infection in Philippine bats", *Virology Journal*, no. 12, p. 107, 2015.

[JEA 08] JEANGÈNE VILMER J.B. *Éthique Animale*, PUF, Paris, 2008.

[JET 17] JETZ W., WILCOVE D.S., DOBSON A.P., "Projected impacts of climate and land-use change on the global diversity of birds", *PLoS Biology*, no. 5, e157, 2017.

[JOH 15] JOHNSON P.T.J., OSTFELD R.S., KEESING F., "Frontiers in research on biodiversity and disease", *Trends in Ecology and Evolution*, vol. 18, pp. 1119–1133, 2015.

[JON 08] JONES K.E., PATEL N.G., LEVY M.A. *et al.*, "Global trends in emerging infectious diseases", *Nature*, no. 451, pp. 990–994, 2008.

[JON 15] JONES R., *House Guests House Pests. A Natural History of Animals in the Home*, Bloomsbury, London, 2015.

[JOS 12] JOSHUA BISHOP (ed.), *The Economics of Ecosystems and Biodiversity in Business and Enterprise*, Earthscan, 2012.

[JØR 10] JØRGENSEN S.E., XU F.-L., COSTANZA R., *Handbook of Ecological Indicators for Assessment of Ecosystem Health*, CRC Press, London, 2010.

[KAH 97] KAHN P., "Developmental psychology and the biophilia hypothesis: children's affiliation with nature", *Developmental Review*, no. 17, pp. 1–61, 1997.

[KAI 14] KAI Z., WOAN T.S., JIE L. *et al.*, "Shifting baselines on a tropical forest frontier: extirpations drive declines in local ecological knowledge", *PLoS ONE*, no. 9, e86598, 2014.

[KAP 89] KAPLAN R., KAPLAN S., *The Experience of Nature: A Psychological Perspective*, Cambridge University Press, Cambridge, 1989.

[KAP 95] KAPLAN S., "The restorative effects of nature: Toward an integrative framework", *Journal of Environmental Psychology*, no. 15, pp. 169–182, 1995.

[KAR 96] KARR J.R., "Ecological integrity and ecological health are not the same", in SCHULZE P.C. (ed.), *Engineering within Ecological Ecologicalecological Constraints*, National Academy Press, Washington, DC, pp. 97–110, 1996.

[KAR 11] KAREIVA P., TALLIS H., RICKETTS T.H. *et al.*, *Natural Capital Theory and Practice of Mapping Ecosystem Services*, Oxford University Press, Oxford, 2011.

[KEE 10] KEESING F., BELDEN L.K., DASZAK P.P. *et al*, "Impacts of biodiversity on the emergence and transmission of infectious diseases", *Nature*, no. 468, pp. 647–652, 2010.

[KÉF 14] KÉFI S., GUTTAL V., BROCK W.A. *et al.*, "Early warning signals of ecological transitions: methods for spatial patterns", *PLoS ONE*, no. 9, e92097, 2014.

[KEL 93] KELLERT R., WILSON E.O., *The Biophilia Hypothesis*, Island Press, Washington, DC, 1993.

[KEL 06] KELLERT S.H., LONGINO H.E., WATERS C.K., *Scientific Pluralism*, University of Minnesota Press, Minneapolis, 2006.

[KEL 11] KELLY D.R., *Yuck!: The Nature and Moral Significance of Disgust*, Massachusetts Institute of Technology, Cambridge, 2011.

[KER 12] KERNOHAN A., *Environmental Ethics: An Interactive Introduction*, Broadview Press, Peterborough, 2012.

[KIM 13] KIM R.E., MACKEY B., "International environmental law as a complex adaptive system", *International Environmental Agreements: Politics, Law and Economics*, no. 14, pp. 5–24, 2013.

[KIN 12] KING K.V., LIVELY C.M., "Does genetic diversity limit disease spread in natural host populations?", *Heredity*, no. 109, pp. 199–203, 2012.

[KOE 01] KOELLA J.C., RESTIF O., "Coevolution of host virulence and host life history", *Ecology Letters*, no. 4, pp. 207–214, 2001.

[KOH 04] KOH L.P., DUNN R.R., SODHI N.S. *et al.*, "Species coextinctions and the biodiversity crisis", *Science*, no. 305, pp. 1632–1634, 2004.

[KRA 09] KRAUT R., *What is Good and Why: The Ethics of Well-Being*, Harvard University Press, Cambridge, 2009.

[KRA 13] KRAUSMANN F., ERB K.-H., GINGRICH S. *et al.*, "Global human appropriation of net primary production doubled in the 20th century", *Proceedings of the National Academy of Sciences USA*, no. 110, pp. 10324–10329, 2013.

[KRI 00] KRISHNAMANI R., MAHANEY W.C., "Geophagy among primates: adaptive significance and ecological consequences", *Animal Behaviour*, no. 59, pp. 899–915, 2000.

[KRI 05] KRIEF S., HLADIK C.M., HAXAIRE C., "Ethnomedicinal and bioactive properties of plants ingested by wild chimpanzees in Uganda", *Journal of Ethnopharmacology*, no. 101, pp. 1–15, 2005.

[KRI 15] KRIEF S., DAUJEARD C., MONCEL M.-H. *et al.*, "Flavouring food: the contribution of chimpanzee behaviour to the understanding of Neanderthal calculus composition and plant use in Neanderthal diets", *Antiquity*, no. 89, pp. 464–471, 2015.

[KRY 70] KRYTER K.D., *The Effects of Noise on Man*, Academic Press, New York, 1970.

[KUH 70] KUHN T.S., *The Structure of Scientific Revolutions*, 2nd ed., University of Chicago Press, Chicago, 1970.

[KUY 94] KUYKEN W., ORLEY J., HUDELSON P. *et al.*, "Quality of life assessment across cultures", *International Journal of Mental Health*, no. 23, pp. 5–27, 1994.

[LAA 11] LAATIKAINEN T., VON HERTZEN L., KOSKINEN J.P. *et al.*, "Allergy gap between Finnish and Russian Karelia on increase", *Allergy*, no. 66, pp. 886–892, 2011.

[LAB 10] LABRIE S.J., SAMSON J.E., MOINEAU S., "Bacteriophage resistance mechanisms", *Nature Review Microbiology*, no. 8, pp. 317–327, 2010.

[LAC 99] LACEY. H., *Is Science Value Free? Values and Scientific Understanding*, Routledge, London, 1999.

[LAF 09] LAFFERTY K.D., "The ecology of climate change and infectious diseases", *Ecology*, no. 90, pp. 888–900, 2009.

[LAF 13] LAFFERTY K.D., WOOD C.L., "It's a myth that protection against disease is a strong and general service of biodiversity conservation: response to Ostfeld and Keesing", *Trends in Ecology & Evolution*, no. 28, pp. 503–504, 2013.

[LAJ 15] LAJAUNIE C., MORAND S., "A legal tool for participatory methods in land systems science: the Thai model of Health Impact Assessment and the consideration of zoonotic diseases concerns into policies", *GLP Newsletter*, no. 11, pp. 30–33, 2015.

[LAJ 16] LAJAUNIE C., MAZZEGA P., "One Health and biodiversity conventions. The emergence of health issues in biodiversity conventions", *IUCN Academy of Environmental Law eJournal*, no. 7, pp. 105–121, 2016.

[LAJ 17a] LAJAUNIE C., MAZZEGA P., "Transmission, circulation et persistance des enjeux de santé dans les conventions internationales liées à la Biodiversité et Conventions de Rio", in MALJEAN-DUBOIS S. (ed.), *Diffusion de normes et circulations d'acteurs dans la gouvernance internationale de l'environnement*, DICE, Toulon, Pau and Aix-en-Provence, 2017.

[LAJ 17b] LAJAUNIE C, MORAND S., "International and regional governance in health and biodiversity", in MORAND S., LAJAUNIE C., ROJCHAI SATRAWAHA R. (eds), *Biodiversity conservation in Southeast Asia: Challenges in a Changing Environment*, Routledge Earthscan, Oxon, 2017.

[LAL 10] LALAND K.N., ODLING-SMEE J., MYLES S., "How culture shaped the human genome: bringing genetic and human sciences together", *Nature Review*, no. 11, pp. 137–145, 2010.

[LAM 17] LAMB J.B., VAN DE WATER J.A.J.M., BOURNE D.G. *et al.*, "Seagrass ecosystems reduce exposure to bacterial pathogens of humans, fishes, and invertebrates", *Science*, no. 355, pp. 731–733, 2017.

[LAN 16] LANGDON A., CROOK N., DANTAS G., "The effects of antibiotics on the microbiome throughout development and alternative approaches for therapeutic modulation", *Genome Medicine*, no. 8, p. 39, 2016.

[LAU 09] LAURANCE W.F., GOOSEM M., LAURANCE S.G.W., "Impacts of roads and linear clearings on tropical forests", *Trends Ecology and Evolution*, no. 24, pp. 659–669, 2009.

[LAW 07] LAWRENCE A., "Beyond the second generation: towards adaptiveness in participatory forest management", *CAB Reviews: Perspectives in Agriculture, Veterinary Science, Nutrition and Natural Resources*, vol. 2, no. 028, p. 1, 2007.

[LAW 07] LAWRENCE A., "No personal motive? Volunteers, biodiversity, and the false dichotomies of participation", *Ethics, Place and Environment*, no. 9, p. 20065, 2007.

[LAW 10] LAWRENCE A., "Introduction: learning from experiences of participatory biodiversity assessment", in LAWRENCE A. (ed.), *Taking Stock of Nature: Participatory Biodiversity Assessment for Policy, Planning and Practice*, Cambridge University Press, Cambridge, 2010.

[LAZ 09] LAZARUS R.J., "Super wicked problems and climate change: restraining the present to liberate the future", *Cornell Law Review*, no. 94, pp. 1153–1234, 2009.

[LEA 10] LEADLEY P., PEREIRA H.M., ALKEMADE R. *et al.*, Biodiversity Scenarios: Projections of 21st century change in biodiversity and associated ecosystem services, Technical Series no. 50, Secretariat of the Convention on Biological Diversity, Montreal, 2010.

[LEC 11] LECHOPIER N., "Sciences, valeurs et pluralisme chez Hugh Lacey", in PIERRON J.-P., PARIZEAU M.-H. (eds), *Nature, Technologie, Ethique. Dialogue Europe, Asie, Amériques*, Les Presses de l'Université Laval, Quebec, 2011.

[LED 92] LEDERBERG L., *Emerging Infections: Microbial Threats to Health in the United States*, National Academy Press, Washington, DC, 1992.

[LEE 15] LEENDERTZ S.A.J., GOGARTEN J.F., DÜX A. *et al.*, "Assessing the evidence supporting fruit bats as the primary reservoirs for Ebola viruses", *Ecohealth*, no. 13, pp. 18–25, 2015.

[LEO 49] LEOPOLD A., *A Land County Almanac and Sketches Here and There*, Oxford University Press, New York, 1949.

[LEV 92] LEVIN S.A., "Orchestrating environmental research assessment", *Ecological Applications*, no. 2, pp. 103–106, 1992.

[LEV 04] LEVIN B.R., BULL J.J., "Population and evolutionary dynamics of phage therapy", *Nature Review Microbiology*, no. 2, pp. 166–173, 2004.

[LEV 12a] LEVENGOOD J., BEASLEY V.R., "Ecotoxicology: bridging wildlife, humans, and ecosystems", in AGUIRRE A.A., OSTFELD R.S., DASZAK P.P. (eds), *New Directions in Conservation Medicine: Applied Cases of Ecological Health*, Oxford University Press, Oxford, 2012.

[LEV 12b] LEVI T., KILPATRICK M.A., MANGEL M. *et al.*, "Deer, predators, and the emergence of Lyme disease", *Proceedings of the National Academy of Sciences USA*, no. 109, pp. 10942–10947, 2012.

[LEV 12c] LEVY K., DAILY G., MYERS S.S., "Human health as an ecosystem service: a conceptual framework", in INGRAM J.C., DECLERCK F., RUMBAITIS DEL RIO C. (eds), *Integrating Ecology and Poverty Reduction: Ecological Dimensions*, Springer Science+Business Media, pp. 231–251, 2012.

[LHO 14] L'HOSTIS M., OLIVIER LAMBERT O., "Comment les pesticides attaquent-ils la santé des pollinisateurs?", in MORAND S., MOUTOU F., RICHOMME C. (eds), *Faune sauvage, biodiversité et santé, quels défis?*, Quae, Versailles, 2014.

[LID 16] LIDDICOAT C., WAYCOTT M., WEINSTEIN P., "Environmental change and human health: can environmental proxies inform the biodiversity hypothesis for protective microbial–human contact?", *BioScience*, no. 66, pp. 1023–1034, 2016.

[LIN 15a] LIN Y., TUHOLSKE J., "Green NGOs win China's first environmental public interest litigation: the Nanping case", *Environmental Law Reporter*, vol. 45, pp. 11102–11103, 2015.

[LIN 15b] LINDAHL J., GRACE D., "The consequences of human actions on risks for infectious diseases: a review", *Infection Ecology Epidemiology*, no. 5, p. 11, 2015.

[LIP 14] LIPSITCH M., GALVANI A.P.P., "Ethical alternatives to experiments with novel potential pandemic pathogens", *PLoS Medicine*, no. 11, p. e1001646, 2014.

[LIS 09] LISONBEE L.D., VILLALBA J.J., PROVENZA F.D. *et al.*, "Tannins and self-medication: implications for sustainable parasite control in herbivores", *Behavioural Processes*, no. 82, pp. 184–189, 2009.

[LIU 10] LIU W., LI Y., LEARN G.H. *et al.*, "Origin of the human malaria parasite Plasmodium falciparum in gorillas", *Nature*, no. 467, pp. 420–425, 2010.

[LIU 15] LIU N., "Insecticide resistance in mosquitoes: impact, mechanisms, and research directions", *Annual Review Entomology*, no. 60, pp. 537–559, 2015.

[LIU 16] LIU Y.-Y., WANG Y., WALSH T.R. *et al.*, "Emergence of plasmid-mediated colistin resistance mechanism MCR-1 in animals and human beings in China: a microbiological and molecular biological study", *Lancet Infectious Diseases*, no. 2, pp. 161–168, 2016.

[LIV 53] LIVADAS G.A., GEORGOPOULOS G., "Development of resistance to DDT by *Anopheles sacharovi* in Greece", *Bulletin of the World Health Organization*, no. 8, pp. 497–511, 1953.

[LLO 09] LLOYD-SMITH J.O., GEORGE D., PEPIN K.M. *et al.*, "Epidemic dynamics at the human-animal interface", *Science*, no. 326, pp. 1362–1367, 2009.

[LOG 03] LOGIUDICE K., OSTFELD R.S., SCHMIDT K.A. *et al.*, "The ecology of infectious disease: effects of host diversity and community composition on Lyme disease risk", *Proceedings of the National Academy of Sciences USA*, no. 100, pp. 567–571, 2003.

[LOO 12] LOOFT T., JOHNSON T.A., ALLEN H.K. *et al.*, "In-feed antibiotic effects on the swine intestinal microbiome", *Proceedings National Academy of Sciences USA*, no. 109, pp. 1691–1696, 2012.

[LOV 95] LOVELOCK J., *The Ages of Gaia: A Biography of Our Living Earth* W.W. Norton & Co, London, 1995.

[LOV 05] LOVEJOY T.E., HANNAH L.J., *Climate Change and Biodiversity*, Yale University Press, Cambridge, 2005.

[LOW 98] LOW N., GLEESON B., *Justice, Society and Nature. An Exploration of Political Ecology*, Routledge, Oxon, 1998.

[LOW 99] LOW N., *Global Ethics and Environment*, Routledge, Oxon, 1999.

[LOZ 98] LOZANO G.A., "Parasitic stress and self-medication in wild animals", *Advances Studies in Behaviour*, no. 27, pp. 291–317, 1998.

[LU 03] LU F., LI Z., "A model of ecosystem health and its application", *Ecological Modelling*, no. 170, pp. 55–59, 2003.

[LU 11] LU T.K., KOERIS M.S., "The next generation of bacteriophage therapy", *Current Opinion in Microbiology*, no. 14, pp. 524–531, 2011.

[LUB 09] LUBY S.P., GURLEY E.S., HOSSAIN M.J., "Transmission of human infection with Nipah virus", *Clinical Infectious Disease*, no. 49, pp. 1743–1748, 2009.

[LUS 12] LUSTIGMAN S., PRICHARD R.K., GAZZINELLI A. *et al.*, "A research agenda for helminth diseases of humans: the problem of helminthiases", *PLoS Neglected Tropical Diseases*, no. 6, p. e1582, 2012.

[LYY 08] LYYTIMÄKI J., PETERSEN L.K., NORMANDER B. *et al.*, "Nature as a nuisance? Ecosystem services and disservices to urban lifestyle", *Environmental Science*, no. 5, pp. 161–172, 2008.

[LYY 14] LYYTIMÄKI J., "Bad nature: newspaper representations of ecosystem disservices", *Urban Forestry and Urban Greening*, no. 13, pp. 418–424, 2014.

[MA 05] MA, *Millenium Ecosystem Assessment. Ecosystems and Human Well–Being: Biodiversity Synthesis*, Island Press, Washington, DC, 2005.

[MAC 04] MACKINNON M.J., READ A.F., "Immunity promotes virulence evolution in a malaria model", *PLoS Biology*, no. 2, p. e230, 2004.

[MAC 08a] MACKEY B., "The earth charter, ethics and global governance", in WENSTRA L., BOSSELMANN K., WESTRA R. (eds), *Reconciling Human Existence with Ecological Integrity. Science, Ethics, Economics and Law*, Earthscan, London, 2008.

[MAC 08b] MACKINNON M.J., GANDON S., READ A.F., "Virulence evolution in response to vaccination: the case of malaria", *Vaccine*, no. 18, pp. C42–C52, 2008.

[MAF 05] MAFFI L., LINGUISTIC, "Cultural and biological diversity?", *Annual Review of Anthropology*, no. 29, pp. 599–617, 2005.

[MAF 08] MAFFI L., "Biocultural diversity and sustainability", in PRETTY J., BALL A., BENTON T. *et al.* (eds), *The SAGE Handbook of Environment and Society*, SAGE Publications, London, 2008.

[MAF 10] MAFFI L., WOODLEY E., *Biocultural Diversity Conservation. A Global Source Book*, Routledge Earthscan, London, 2010.

[MAI 12] MAIZELS R.M., HEWITSON J.P., SMITH K.A., "Susceptibility and immunity to helminth parasites", *Current Opinion Immunology*, no. 24, pp. 459–466, 2012.

[MAI 16] MAIZELS R.M., MCSORLEY H.J., "Regulation of the host immune system by helminth parasites", *Journal of Allergy and Clinical Immunology*, no. 138, pp. 666–675, 2016.

[MAL 06] MALCOLM J.R., LIU C., NEILSON R.P. *et al.*, "Global warming and extinctions of endemic species from biodiversity hotspots", *Conservation Biology*, no. 20, pp. 538–548, 2006.

[MAN 07] MANZO L.C., BRIGHTBILL B., "Toward a participatory ethics", in KINDON S., PAIN R., KESBY M. (eds), *Participatory Action Research Approaches and Methods: Connecting People, Participation, and Place*, Routledge, Oxon, pp. 33–40, 2007.

[MAN 15] MANCISIDOR M., "Is there such a thing as a human right to science in international law?", *European Society of International Law Reflections*, no. 4, p. 6, 2015.

[MAR 09a] MARTENSEN R., "Landscape designers, doctors, and the making of healthy urban spaces in 19th century America", in CAMPBELL L., WIESEN A. (eds), *Restorative Commons: Creating Health and Well-being Through Urban Landscapes*, General Technical Report NRS-P-39, U.S. Department of Agriculture, Forest Service, Northern Research Station, pp. 26–37, 2009.

[MAR 09b] MARTIN L.B., "Stress and immunity in wild vertebrates: timing is everything", *General and Comparative Endocrinology*, no. 163, pp. 70–76, 2009.

[MAR 09c] MARTINEZ J.L., "The role of natural environments in the evolution of resistance traits in pathogenic bacteria", *Proceedings Royal Society London B*, no. 276, pp. 2521–2530, 2009.

[MAR 10a] MARRIS V., *Philosophie de la biodiversité: Petite éthique pour une nature en péril*, Buchet-Chastel, Arles, 2010.

[MAR 10b] MARTENS P., AKIN S.-M., MAUD H. *et al.*, "Is globalization healthy: a statistical indicator analysis of the impacts of globalization on health", *Global Health*, no. 6, p. 16, 2010.

[MAR 14] MARRIS V., *Nature à Vendre. Les limites des Services Écosystémiques*, Quae, Versailles, 2014.

[MAR 15] MARTINEZ-JUAREZ P., CHIABAI A., TAYLOR T. *et al.*, "The impact of ecosystems on human health and well-being: a critical review", *Journal of Outdoor Recreation and Tourism*, no. 10, pp. 63–69, 2015.

[MAR 16] MARSTON H.D., DIXON D.M., KNISELY J.M. *et al.*, "Antimicrobial resistance", *Journal of the American Medical Association*, no. 311, pp. 1193–204, 2016.

[MAT 06] MATHERS C.D., LONCAR D., "Projections of global mortality and burden of disease from 2002 to 2030", *PLoS Medicine*, no. 3, pp. 2011–2030, 2006.

[MAT 12] MATHEVET R., *La Solidarité Ecologique*, Actes Sud, Arles, 2012.

[MAY 50] MAY J.M., "Medical geography: its methods and objectives", *Geographical Review*, no. 40, pp. 9–41, 1950.

[MAY 52] MAY J.M., History, definition and progress of medical geography: a general review, First Report of the Commission on Medical Geography of the International Geographical Union, IGU, London, pp. 1–9, 1952.

[MAY 58] MAY J.M., *The Ecology of Human Disease*, MD Publications, New York, 1958.

[MAY 15] MAYNARD J., VAN HOOIDONK R., MARK EAKIN C.M. *et al.*, "Projections of climate conditions that increase coral disease susceptibility and pathogen abundance and virulence", *Nature Climate Change*, no. 5, pp. 688–694, 2015.

[MCC 13] MCCLURE C.J.W., WARE H.E., CARLISLE J. *et al.*, "An experimental investigation into the effects of traffic noise on distributions of birds: avoiding the phantom road", *Proceedings of the Royal Society of London B*, no. 280, p. 20132290, 2013.

[MCD 07] MCDADE T.W., REYES-GARCÍA V., BLACKINTON P. *et al.*, "Ethnobotanical knowledge is associated with indices of child health in the Bolivian Amazon", *Proceedings National Academy of Sciences USA*, no. 104, pp. 6134–6139, 2007.

[MCD 13] MCDONALD R.I., MARCOTULLIO P.J., BURAK GÜNERALP B. *et al.*, "Urbanization and global trends in biodiversity and ecosystem services", in ELMQVIST T., FRAGKIAS M., GOODNESS J. *et al.* (eds), *Urbanization, Biodiversity and Ecosystem Services: Challenges and Opportunities: A Global Assessment*, Springer, Dordrecht, 2013.

[MCG 99] MCGINNIS M.V., *Bioregionalism*, Routledge, Oxon, 1999.

[MCG 06] MCGRANAHAN G., MARCOTULLIO P., BAI X. *et al.*, "Urban systems", in HASSAN R., SCHOLES R., ASH N. (eds), *Ecosystems and Human Well-Being: Current State and Trends*, vol. 1, Island Press, Washington, DC, 2006.

[MCG 07] MCGILLIVRAY M., *Human Well-Being: Concept And Measurement*, Palgrave Macmillan, London, 2007.

[MCN 76] MCNEIL W.H., *Plagues and People*, Anchor Press, New York, 1976.

[MCN 00] MCNEILL J.R., *Something New Under the Sun: An Environmental History of the 20th-Century World*, W.W. Norton, New York, 2000.

[MEA 72] MEADOWS D.H., MEADOWS D.L., RANDERS J. *et al.*, *The Limits to Growth: A Report for the Club of Rome's Project on the Predicament of Mankind*, Universe Books, 1972.

[MEA 00] MEADE M., EARICKSON R., *Medical Geography*, 2nd ed., Guilford Press, New York, 2000.

[MEA 06] MEADOWS D.H., RANDERS J., MEADOWS D.L., *Limits to Growth: The 30-Year Update*, EarthScan, London, 2006.

[MEE 09] MEERBURG B.G., SINGLETON G.R., KIJLSTRA A., "Rodent-borne diseases and their risks for public health", *Critical Review in Microbiology*, no. 35, pp. 221–270, 2009.

[MEE 14] MEERBURG B.G., VAN GENT-PELZER M.P.P.E., SCHOELITSZ B. *et al.*, "Distribution of anticoagulant rodenticide resistance in *Rattus norvegicus* in the Netherlands according to Vkorc1 mutations", *Pest Management Science*, no. 70, pp. 1761–1766, 2014.

[MEN 95] MENDELSOHN R., BALICK M.J., "The value of undiscovered pharmaceuticals in tropical forests", *Economic Botany*, no. 49, pp. 223–228, 1995.

[MEN 03] MENDELSOHN M., KOUGH J., VAITUZIS Z. *et al.*, "Are Bt crops safe?", *Nature Biotechnology*, no. 21, pp. 1003–1009, 2003.

[MIL 08] MILLER T., BAIRD T.D., LITTLEFIELD C.M. *et al.*, "Epistemological pluralism: reorganizing interdisciplinary research", *Ecology and Society*, no. 13, p. 46, 2008.

[MIL 09] MILLER T., SPOOLMAN S.E., *Sustaining the Earth*, 10th ed., Brooks/Cole, Pacific Grove, 2009.

[MIR 16] MIRALDO A., LI S., BORREGAARD M.K. *et al.*, "An Anthropocene map of genetic diversity", *Science*, no. 353, pp. 1532–1535, 2016.

[MOH 07] MOHAN G., "Participatory development: from epistemological reversals to active citizenship", *Geography Compass*, no. 1, pp. 779–796, 2007.

[MON 09] MONOT M., HONORÉ N., GARNIER T. *et al.*, "Comparative genomic and phylogeographic analysis of *Mycobacterium leprae*", *Nature Genetics*, no. 41, pp. 1282–1289, 2009.

[MOO 15] MOORE J.W., *Capitalism in the Web of Life: Ecology and the Accumulation of Capital*, Verso, London, 2015.

[MOR 90] MORSE S.S., SCHLUEDERBERG A., "Emerging viruses: the evolution of viruses and viral diseases", *The Journal of Infectious Diseases*, no. 162, pp. 1–7, 1990.

[MOR 95] MORSE S.S., "Factors in the emergence of infectious diseases", *Emerging Infectious Diseases*, no. 1, pp. 7–15, 1995.

[MOR 08a] MORAND S., DETER J., "Parasitism and regulation of the host population", in THOMAS F., GUÉGAN J.F., RENAUD F. (eds), *Ecology and Evolution of Parasitism*, Oxford University Press, Oxford, 2008.

[MOR 08b] MORT M., CONVERY I., BAXTER J. *et al.*, "Animal disease and human trauma: the psychosocial implications of the 2001 UK Foot and Mouth disease disaster", *Journal of Applied Animal Welfare Science*, no. 11, pp. 133–148, 2008.

[MOR 10a] MORGERA E., TSIOUMANI E., "The evolution of benefit sharing: linking biodiversity and community livelihoods", *Review of European Community & International Environmental Law*, no. 19, pp. 150–173, 2010.

[MOR 10b] MORRIS R.J., "Anthropogenic impacts on tropical forest biodiversity: a network structure and ecosystem functioning perspective", *Philosophical Transactions of the Royal Society London B*, no. 365, pp. 3709–3718, 2010.

[MOR 11a] MORAND S., "Infectious diseases, biodiversity and global changes: how the biodiversity sciences may help", in LOPEZ PUJOL J. (ed.), *Biodiversity/Book 3*, Intech, 2011.

[MOR 11b] MORAND S., "La biodiversité comme dimension du changement global", *Recherches Internationales*, no. 89, pp. 213–230, 2011.

[MOR 12a] MORAND S., "Phylogeography helps with investigating the building of human parasite communities", *Parasitology*, no. 139, pp. 1966–1974, 2012.

[MOR 12b] MORAND S., WARET-SZKUTA A., "Determinants of human infectious diseases in Europe: biodiversity and climate variability influences", *Bulletin Epidémiologique Hebdomadaire*, no. 12–13, pp. 156–159, 2012.

[MOR 13a] MORAND S., OWERS K., WARET-SZKUTA A. *et al.*, "Climate variability and outbreaks of infectious diseases in Europe", *Scientific Reports*, no. 3, p. 1774, 2013.

[MOR 13b] MORRISON V., Les Problèmes Vicieux Et Les Politiques Publiques, Centre de Collaboration Nationale sur les Politiques Publiques et la Santé, Montreal, Quebec, 2013.

[MOR 14a] MORAND S., "Impact of climate change on livestock disease occurrences", in SEJIAN V. (ed.), *Climate change Impact on Livestock: Adaptation and Mitigation*, Springer, New-York, 2014.

[MOR 14b] MORAND S., JITTAPALAPONG S., SUPPUTAMONGKOL Y. *et al.*, "Infectious diseases and their outbreaks in Asia-Pacific: biodiversity and its regulation loss matter", *PLoS One*, no. 9, p. e90032, 2014.

[MOR 14c] MORAND S., MCINTYRE K.M., BAYLIS M., "Domesticated animals and human infectious diseases of zoonotic origins: domestication time matters", *Infection Genetics Evolution*, no. 24, pp. 76–87, 2014.

[MOR 14d] MORAND S., OWERS K., BORDES F., "Biodiversity and emerging zoonoses", in YAMADA A., KAHN L.H., KAPLAN B. *et al.* (eds), *Confronting Emerging Zoonoses: The One Health Paradigm*, Springer, Tokyo, 2014.

[MOR 14e] MORGERA E., TSIOUMANI E., BUCK M., *Unraveling the Nagoya Protocol: a commentary on the Nagoya Protocol on Access and Benefit-sharing to the Convention on Biological Diversity*, Brill/Martinus Nijhoff, 2014.

[MOR 15a] MORAND S., "Diversity and origins of human infectious diseases", in MUEHLENBEIN M.P. (ed.), *Basics in Human Evolution*, Elsevier, New York, 2015.

[MOR 15b] MOREAU P., FAURY N., BURGEOT T. *et al.*, "Pesticides and ostreid herpesvirus 1 infection in the Pacific oyster, *Crassostrea gigas*", *PLoS One*, no. 10, pp. 1–12, 2015.

[MOR 15c] MORGERA E., "Fair and equitable benefit-sharing at the cross-roads of the human right to science and international biodiversity law", *Laws*, no. 4, pp. 803–831, 2015.

[MOR 16] MORAND S., *La Prochaine Peste, Une Histoire Globale des Maladies Infectieuses*, Fayard, Paris, 2016.

[MOR 17a] MORIN E., *Connaissance, Ignorance, Mystère*, Fayard, Paris, 2017.

[MOR 17b] MORAND S., "Biodiversity and disease transmission", in HURST C.J. (ed.), *The Connections between Ecology and Infectious Disease*, Advances in Environmental Microbiology, Springer, 2017.

[MOR 17c] MORAND S, LAJAUNIE C., "The History and Diversity of South-east Asia", in MORAND S., LAJAUNIE C., SATRAWAHA R. (eds), *Biodiversity Conservation in Southeast Asia: Challenges in a Changing Environment*, Routledge EarthScan, Oxon, 2017.

[MOT 14] MOTESHARREI S., RIVAS J., KALNAY E. *et al.*, "Human and nature dynamics (HANDY): modeling inequality and use of resources in the collapse or sustainability of societies", *Ecological Economics*, no. 101, pp. 90–102, 2014.

[MOU 15] MOUQUET N., LAGADEUC Y., DEVICTOR V. *et al.*, "Predictive ecology in a changing world", *Journal Applied Ecology*, no. 52, pp. 1293–1310, 2015.

[MUE 05] MUERHOFF A.S., LEARY T.P., SATHAR M.A. *et al.*, "African origin of GB virus C determined by phylogenetic analysis of a complete genotype 5 genome from South Africa", *Journal of General Virology*, no. 86, pp. 1729–735, 2005.

[MUE 10] MUEHLENBEIN M.M., "Evolutionary medicine, immunity, and infectious disease", in MUEHLENBEIN M.M. (ed.), *Human Evolutionary Biology*, Cambridge University Press, Cambridge, pp. 459–490, 2010.

[MUR 13] MURRAY K.A., DASZAK P., "Human ecology in pathogenic landscapes: two hypotheses on how land use change drives viral emergence", *Current Opinion in Virology*, no. 3, pp. 79–83, 2013.

[MYE 00] MYERS N., MITTERMEIER R.A., MITTERMEIER C.G. *et al.*, "Biodiversity hotspots for conservation priorities", *Nature*, no. 403, pp. 853–858, 2000.

[MYE 13] MYERS S.S., GAFFIKIN L., GOLDEN C.D. *et al.*, "Human health impacts of ecosystem alteration", *Proceedings of the National Academy of Sciences USA*, no. 110, pp. 18753–18760, 2013.

[NAE 89] NAESS A., *Ecology, Community and Lifestyle*, Cambridge University Press, Cambridge, 1989.

[NAE 16] NAEEM S., CHAZDON R., DUFFY J.E. *et al.*, "Biodiversity and human well-being: an essential link for sustainable development", *Proceedings of the Royal Society B*, no. 283, p. 20162091, 2016.

[NAG 13] NAGENDRA H., REYERS B., LAVOREL S., "Impacts of land change on biodiversity: making the link to ecosystem services", *Current Opinion in Environmental Sustainability*, no. 5, pp. 503–508, 2013.

[NAS 89] NASH R., *The Rights of Nature*, The University of Wisconsin Press, Madison, 1989.

[NAS 05] NAS, *Valuing Ecosystem Services: Toward Better Environmental Decision-Making*, National Academic Press, Washington, DC, 2005.

[NAS 06] NASH L., *Inescapable Ecologies. A History of Environment, Diseases, and Knowledge*, University of California Press, 2006.

[NAS 17] NAS, *Communicating Science Effectively: A Research Agenda*, National Academy of Sciences, Washington, DC, 2017.

[NAV 07] NAVARA K.J., NELSON R.J., "The dark side of light at night: physiological, epidemiological, and ecological consequences", *Journal of Pineal Research*, no. 43, pp. 215–224, 2007.

[NEL 10] NELSON, E., SANDER, H., HAWTHORNE, P. *et al.*, "Projecting global land-use change and its effect on ecosystem service provision and biodiversity with simple models", *PLoS ONE*, no. 5, p. e14327, 2010.

[NES 96] NESSE R.M., WILLIAMS G.C., *Why We Get Sick: the New Science of Darwinian Medicine*, Vintage Books, New York, 1996.

[NEW 11] NEWING H., *Conducting Research in Conservation Social Science.Methods and Practice*, Routledge, Oxon, 2011.

[NGO 08] NGOC VAN N.D., TAP N., An Overview of the Use of Plants and Animals in Traditional Medicine Systems in Viet Nam, TRAFFIC Southeast Asia, Greater Mekong Programme, Ha Noi, Vietnam, 2008.

[NHU 15] NHUNG N.T., CUONG N.V., CAMPBELL J. *et al.*, "High levels of antimicrobial resistance among *Escherichia coli* isolates from livestock farms and synanthropic rats and shrews in the Mekong Delta of Vietnam", *Applied Environmental Microbiology*, no. 81, pp. 812–820, 2015.

[NIC 08] NICHTER M., *Global Health: Why Cultural Perceptions, Social Representations, and Biopolitics Matter*, University of Arizona Press, 2008.

[NOL 11] NOLAN J.M., TURNER N.J., "Ethnobotany: the study of people–plant relationships", in ANDERSON E.N., PEARSALL D., HUNN E. *et al.* (eds), *Ethnobiology*, John Wiley & Sons, London, pp. 133–147, 2011.

[NOR 06] NORRIS D.O., CARR J.A., *Endocrine Disruption: Biological Basis for Health Effects in Wildlife and Humans*, Oxford University Press, Oxford, 2006.

[NUN 10] NUNN N., QIAN N., "The Columbian exchange: a history of disease, food, and ideas", *Journal of Economic Perspectives*, no. 24, pp. 163–188, 2010.

[OBR 10] O'BRIEN K.J., *An Ethics of Biodiversity: Christianity, Ecology, and the Variety of Life*, Georgetown University Press, Washington, DC, 2010.

[OEC 15] OECD, Alternative futures for global food and agriculture: developing robust strategies, Joint Working Party On Agriculture And Trade, TAD/TC/CA/WP, available at: http://www.oecd.org/fr/publications/ alternative-futures-for-global-food-and-agriculture-9789264247826-en.htm, 2015.

[OFA 11] O'FALLON B.D., FEHREN-SCHMITZ L., "Native Americans experienced a strong population bottleneck coincident with European contact", *Proceedings of the National Academy of Sciences USA*, no. 108, pp. 20444–20448, 2011.

[ÖHR 06] ÖHRSTRÖM E., HADZIBAJRAMOVIC E., HOLMES M. *et al.*, "Effects of road traffic noise on sleep: studies on children and adults", *Journal of Environmental Psychology*, no. 26, pp. 116–126, 2006.

[OLA 15] OLAITAN A., THONGMALAYVONG B., AKKHAVONG K. *et al.*, "Clonal transmission of a colistin-resistant *Escherichia coli* from a domesticated pig to human in Laos", *Journal Antimicrobical Chemistry*, no. 70, pp. 3402–3404, 2015.

[OLA 16] OLAITAN A., CHABOU S., OKDAH L. *et al.*, "Dissemination of the mcr-1 colistin resistance gene", *Lancet Infectious Diseases*, no. 16, pp. 147–149, 2016.

[ONE 14] O'NEILL J., *Antimicrobial Resistance: Tackling a Crisis for the Health and Wealth of Nations*, Wellcome Trust, London, 2014.

[OPP 10] OPPONG J.R., HAROLD A., "Disease, ecology, and environment", in BROWN T., MCLAFFERTY S., MOON G. (eds), *A Companion to Health and Medical Geography*, Wiley, London, 2010.

[ORI 80] ORIANS G.H., "Habitat selection: general theory and applications to human behavior", in LOCKARD J.S. (eds), *The Evolution of Human Social Behavior*, Elsevier North Holland, New York, 1980.

[ORR 93] ORR D.W., "Love it or lose it: the Biophilia revolution", in KELLERT S.R., WILSON E.O. (eds), *The Biophilia Hypothesis*, Island Press, Washington, DC, 1993.

[ORR 11] ORROCK J.L., ALLAN B.F., DROST C.A., "Biogeographic and ecological regulation of disease: prevalence of Sin Nombre virus in island mice is related to island area, precipitation, and predator richness", *American Naturalist*, no. 177, pp. 691–697, 2011.

[OST 90] OSTROM E., *Governing the Commons: The Evolution of Institutions for Collective Action*, Cambridge University Press, Cambridge, 1990.

[OUY 11] OUYANG J.Q., SHARP P., DAWSON A. *et al.*, "Hormone levels predict individual differences in reproductive success in a passerine bird", *Proceedings of the Royal Society London B*, no. 278, pp. 2537–2545, 2011.

[OZO 94] OZOLINS G., STOBER J., WHO Global Strategy for Health and Environment and Related Events, WHO/EOS/94.41, Report, Geneva, 1994.

[PAA 05] PAAVOLA J., LOWE I., *Environmental Values in a Globalising World. Nature, Justice and Governance*, Routledge, London, 2005.

[PAL 01] PALUMBI S.R., "Humans as the world's greatest evolutionary force", *Science*, no. 293, pp. 1786–1790, 2001.

[PAL 16] PAL C., BENGTSSON-PALME J., KRISTIANSSON E. *et al.*, "The structure and diversity of human, animal and environmental resistomes", *Microbiome*, no. 4, pp. 1–15, 2016.

[PAR 11] PARKER B.J., BARRIBEAU S.M., LAUGHTON A.M. *et al.*, "Non-immunological defense in an evolutionary framework", *Trends in Ecology and Evolution*, no. 26, pp. 242–248, 2011.

[PAS 80] PASSMORE J.A., *Man's Responsibility For Nature: Ecological Problems And Western Traditions*, Duckworth, London, 1980.

[PAT 04] PATZ J.A., DASZAK P., TABOR G.M. *et al.*, "Unhealthy landscapes: policy recommendations on land use change and infectious disease emergence", *Environmental Health Perspectives,* no. 112, pp. 1092–1098, 2004.

[PEC 17] PECL G.T., ARAÚJO M.B., BELL J.D. *et al.*, "Biodiversity redistribution under climate change: impacts on ecosystems and human well-being", *Science*, no. 355, p. 1389, 2017.

[PED 15] PEDRONO M., LOCATELLI B., EZZINE-DE-BLAS D. *et al.*, "Impact of climate change on ecosystem services", in TORQUEBIAU E. (ed.), *Climate Change and Agriculture Worldwide*, Quae & Springer, 2015.

[PEL 10] PELLETIER N., TYEDMERS P., "Forecasting potential global environmental costs of livestock production 2000–2050", *Proceedings National Academy of Sciences USA*, no. 107, pp. 18371–18374, 2010.

[PER 13] PERRY B.D., GRACE D., SONES K., "Current drivers and future directions of global livestock disease dynamics", *Proceedings of the National Academy Sciences USA*, no. 110, pp. 20871–20877, 2013.

[PET 01] PETERSON A.L., *Being human: Ethics, Environment, and our Place in the World*, University of California Press, Berkeley 2001.

[PET 14] PETERSON A.T., *Mapping Disease Transmission Risk: Enriching Models Using Biogeography and Ecology*, Johns Hopkins University Press, Baltimore, 2014.

[PHI 01] PHILLIPSON J.D., "Phytochemistry and medicinal plants", *Phytochemistry*, no. 56, pp. 237–245, 2001.

[PHI 11] PHILIPPOPOULOS-MIHALOPOULOS A., *Law and Ecology: New Environmental Foundations*, Routledge, Oxon, 2011.

[PIC 00] PICHERSKY E., GANG D.R., "Genetics and biochemistry of secondary metabolites in plants: an evolutionary perspective", *Trends in Plant Science*, no. 5, pp. 439–445, 2000.

[PIE 07] PIELKE R., *The Honest Broker: Making Sense of Science in Policy and Politics*, Cambridge University Press, Cambridge, 2007.

[PIL 14] PILOSOF S., FORTUNA M.A., COSSON J.-F. *et al.*, "Host–parasite network structure is associated with community-level immunogenetic diversity", *Nature Communications*, no. 5, p. 5172, 2014.

[PIM 84] PIMM S.L., "The complexity and stability of ecosystems", *Nature*, no. 307, pp. 321–326, 1984.

[PLU 16] PLUTZAR C., ERB K.-H., GAUBE V. *et al.*, "Of birds and bees: biodiversity and the colonization of ecosystems", in HABERL H., FISCHER-KOWALSKI M., KRAUSMANN F. *et al.* (eds), *Social Ecology Society-Nature Relations across Time and Space*, Springer International Publishing, Switzerland, 2016.

[POI 15] POISOT T., NUNN C., MORAND S., "Ongoing worldwide homogenization of human pathogens", *BioRxiv*, doi: http://dx.doi.org/10.1101/009977, 2015.

[POP 12] POPPER A.N., HAWKINS A., *The Effects of Noise on Aquatic Life*, Springer, 2012.

[PRI 10] PRING G., & PRING C., "Specialized environmental courts and tribunals: the explosion of new institutions to adjudicate environment, climate change, and sustainable development", *2nd Global Conference on Environmental Governance and Democracy*, available at: http://conference.unitar.org/yale/sites/conference.unitar.org.yale/files/Pring_Paper.pdf, 2010.

[PRI 11] PRIEUR M., "Instruments internationaux et évaluation environnementale de la biodiversité : enjeux et obstacles", *Revue juridique de l'environnement*, vol. spécial, no. 5, pp. 7–28, 2011.

[PRI 89] PRING G., "SLAPPs: strategic lawsuits against public participation", *7 Pace Environmental Law Review*, vol. 7, no. 1, pp. 3–21, 1989.

[PRI 15] PRICE L.B., KOCH B.J., HUNGATE B.A., "Ominous projections for global antibiotic use in food-animal production", *Proceedings National Academy of Sciences USA*, no. 112, pp. 5554–5555, 2015.

[PRI 16] PRING G., PRING C., Environmental Courts and Tribunals: A Guide for Policy-Makers, Report, United Nations Environment Programme, 2016.

[PRI 17] PRING G., PRING C., "Specialized environmental courts and tribunals: the explosion of new institutions to adjudicate environment, climate change, and sustainable development", *2nd Global Conference on Environmental Governance*, available at: http://conference.unitar.org/yale/sites/conference.unitar.org.yale/files/Pring_Paper.pdf, accessed 1 June 2017.

[PRO 16] PROENÇA V., MARTIN L.J., PEREIRA H.M. *et al.*, "Global biodiversity monitoring: from data sources to essential biodiversity variables", *Biological Conservation*, doi.org/10.1016/j.biocon.2016.07.014, 2016.

[PRU 05] PRUGNOLLE F., MANICA A., CHARPENTIER M. *et al.*, "Pathogen-driven selection and worldwide HLA class I diversity", *Current Biology*, no. 15, pp. 1022–1027, 2005.

[PUK 94] PUKRITTAYAKAMEE S., SUPANARANOND W., LOOAREESUWAN S. *et al.*, "Quinine in severe *falciparum* malaria: evidence of declining efficacy in Thailand", *Transactions Royal Society of Tropical Medicine and Hygiene*, no. 88, pp. 324–327, 1994.

[RAB 13] RABINOWITZ P., CONTI L., "Links among human health, animal health, and ecosystem health", *Annual Review of Public Health*, no. 34, pp. 189–204, 2013.

[RAG 13a] RAGHUNATHAN N., ANSAY F., NSENGA L., "Gestion participative des ressources naturelles dans les réserves de biosphère. L'expérience du WWF dans les réserves de Luki, Yangambi (RDC) et Dimonika (République du Congo)", WWF Guide Technique, 2013.

[RAG 13b] RAGON P., "Le médicament dans l'environnement", in MORAND S., PIPIEN G. (eds), *Notre Santé et la Biodiversité*, Buchet-Chastel, Arles, 2013.

[RAM 07] RAMIREZ C.R., "Ethnobotany and the loss of traditional knowledge in the 21st century", *Ethnobotany Research Applications*, no. 5, pp. 245–247, 2007.

[RAN 51] RANKOVIC A., PACTEAU C., ABBADIE L., "Services écosystémiques et adaptation urbaine interscalaire au changement climatique: un essai d'articulation", *VertigO*, special edition, no. 12, available at: https://vertigo.revues.org/11851#quotation, doi: 10.4000/vertigo.11851.

[RAN 16] RANSON H., LISSENDEN N., "Insecticide resistance in African *Anopheles* mosquitoes: a worsening situation that needs urgent action to maintain malaria control", *Trends in Parasitology*, no. 32, pp. 187–196, 2016.

[RAP 95] RAPPORT D.J., GAUDET C.L., CALOW P., *Evaluating and Monitoring the Health of Large-Scale Ecosystems*, Springer-Verlag, Berlin, 1995.

[RAP 98] RAPPORT D.J., COSTANZA R., MCMICHAEL A.J., "Assessing ecosystem health", *Trends in Ecology and Evolution*, no. 13, pp. 397–402, 1998.

[RAP 03] RAPPORT D.J., LASLEY W.L., ROLSTON D.E. *et al.*, *Managing for Healthy Ecosystems*, Lewis Publisher, Boca Raton, 2003.

[RAS 11] RASMUSSEN M., GUO X., WANG Y. *et al.*, "An aboriginal Australian genome reveals separate human dispersals into Asia", *Science*, no. 334, pp. 94–98, 2011.

[RAV 06] RAVI RAJAN S., *Modernizing Nature: Forestry and Imperial Eco-development 1800–1950*, Oxford University Press, Oxford, 2006.

[RAW 71] RAWLS J., *A Theory of Justice*, Harvard University Press, Cambridge, 1971.

[REA 01] READ A.F., TAYLOR L.H., "The ecology of genetically diverse infections", *Science*, no. 292, pp. 1099–1102, 2001.

[REE 96] REED E.S., *Encountering The World: Toward An Ecological Psychology*, Oxford University Press, Oxford, 1996.

[REE 07] REED D.L., LIGHT J.E., ALLEN J.M. *et al.*, "Pair of lice lost or parasites regained: the evolutionary history of anthropoid primate lice", *BMC Biology*, no. 5, p. 7, 2007.

[REG 03] REGAN T., *Animal Rights, Human Wrongs: An Introduction to Moral Philosophy*, Rowman & Littlefield, Lanham, 2003.

[REI 13] REINHARD K.J., FERREIRA L.F., BOUCHET F. *et al.*, "Food, parasites, and epidemiological transitions: a broad perspective", *International Journal of Paleopathology*, no. 3, pp. 150–157, 2013.

[RIT 73] RITTEL H.W.J., WEBBER M.M., "Dilemmas in a general theory of planning", *Policy Sciences*, no. 4, pp. 155–169, 1973.

[RIV 13] RIVASSEAU JONVEAUX T., FESCHAREK R., "Environnement et santé psychologique: des dimensions thérapeutiques de la nature", in MORAND S, PIPIEN G. (eds), *Notre Santé et la Biodiversité*, Buchet-Chastel, Arles, 2013.

[ROA 03] ROACH C.M., *Mother/Nature: Popular Culture and Environmental Ethics*, Indiana University Press, Bloomington, IN, 2003.

[ROB 03] ROBINSON N., "Challenges confronting the progressive development of a second generation of environmental laws", in LYE L.-H., MANGUIAT M., SOCORRO Z. (eds), *Towards a "Second Generation" in Environmental Laws in the Asian and Pacific Region: Select Trends*, IUCN, Gland & Cambridge, 2003.

[ROC 09] ROCKSTRÖM J., STEFFEN W., NOONE K. *et al.*, "Planetary boundaries:exploring the safe operating space for humanity", *Ecology and Society*, no. 14, p. 32, 2009.

[ROH 05] ROHDE K., *Nonequilibrium Ecology*, Cambridge University Press, Cambridge, 2005.

[ROH 06] ROHR J.R., KERBY J.L., SIH A., "Community ecology as a framework for predicting contaminant effects", *Trends in Ecology and Evolution*, no. 21, pp. 606–613, 2006.

[ROL 92] ROLSTON III H., "Challenges in environmental ethics", in COOPER D.E., PALMER J.A. (ed.), *The Environment in Question Ethics and Global Issues*, Routledge, London, pp. 134–144, 1992.

[ROL 12] ROLSTON III H., *A New Environmental Ethics: The Next Millennium for Life on Earth*, Routledge, Oxon, 2012.

[ROM 04] ROMERO L.M., "Physiological stress in ecology: lessons from biomedical research", *Trends in Ecology and Evolution*, no. 9, pp. 249–255, 2004.

[ROO 13] ROOK G.A., "Regulation of the immune system by biodiversity from the natural environment: an ecosystem service essential to health", *Proceedings of the National Academy of Sciences USA*, no. 110, pp. 18360–18367, 2013.

[ROS 09] ROSENTHAL B.M., "How has agriculture influenced the geography and genetics of animal parasites?", *Trends in Parasitology*, no. 25, pp. 67–70, 2009.

[ROS 11] ROSE G.L., "Gaps in the implementation of environmental law at the national, regional and global level", *First Preparatory Meeting of the World Congress on Justice, Governance and Law for Environmental Sustainability*, UNEP, Kuala Lumpur, Malaysia, 12–13 October 2011.

[ROW 99] ROWE G., WRIGHT G., "The Delphi technique as a forecasting tool: issues and analysis", *International Journal of Forecasting*, no. 15, pp. 353–375, 1999.

[ROZ 15a] ROZZI R., CHAPIN III F.S., CALLICOTT J.B. *et al.*, *Earth Stewardshipp. Linking Ecology and Ethics in Theory and Practice*, Springer, London, 2015.

[ROZ 15b] ROZZI R., CHAPIN III F.S., CALLICOTT J.B. *et al.*, "Linking ecology and ethics for an interregional and intercultural earth stewardship", in ROZZI R. *et al.* (eds), *Earth Stewardshipp. Linking Ecology and Ethics in Theory and Practice*, Springer, London, 2015.

[RUD 09] RUDEL T.K., SCHNEIDER L., URIARTE M. *et al.*, "Agricultural intensification and changes in cultivated areas, 1970–2005", *Proceedings of National Academy of Science USA*, no. 106, pp. 20675–20680, 2009.

[RUH 07] RUHL J.B., STEVEN E., KRAFT S.E. *et al.*, *The Law and Policy of Ecosystem Services*, Island Press, Washington, DC, 2007.

[RUS 03] RUSO B., RENNINGER L., ATZWANGER K., "Human habitat preferences: a generative territory for evolutionary aesthetics research", in ECKART VOLAND E., KARL GRAMMER K. (eds), *Evolutionary Aesthetics*, Springer-Verlag, Berlin, 2003.

[SAL 09] SALA O.E., MEYERSON L.A., PARMESAN C., *Biodiversity Change and Human Health: From Ecosystem Services to Spread of Disease*, Island Press, Washington, DC, 2009.

[SAL 13] SALKELD D.J., PADGETT K., JONES J.H., "A meta-analysis suggesting that the relationship between biodiversity and risk of zoonotic pathogen transmission is idiosyncratic", *Ecology Letters*, no. 16, pp. 679–686, 2013.

[SAL 15] SALGOTRA R.K., GUPTA B.G., *Plant Genetic Resources and Traditional Knowledge for Food Security*, Springer, Singapore, 2015.

[SAN 11] SANAHUJA G., BANAKAR R., TWYMAN R. *et al.*, "*Bacillus thuringiensis*: a century of research, development and commercial applications", *Plant Biotechnology Journal*, no. 9, pp. 283–300, 2011.

[SAT 10] SATTERTHWAITE D., MCGRANAHAN G., TACOLI C., "Urbanization and its implications for food and farming", *Philosophical Transactions of the Royal Society London B*, no. 365, pp. 2809–2820, 2010.

[SCH 01] SCHMIDT K.A., OSTFELD R.S., "Biodiversity and the dilution effect in disease ecology", *Ecology*, no. 82, pp. 609–619, 2001.

[SCH 08] SCHIPPER J., CHANSON J.S., CHIOZZA F. *et al.*, "The status of the world's land and marine mammals: diversity, threat, and knowledge", *Science*, no. 322, pp. 225–230, 2008.

[SCH 15a] SCHAUB B., VERCELLI D., "Environmental protection from allergic diseases: from humans to mice and back", *Current Opinion Immunology*, no. 36, pp. 88–93, 2015.

[SCH 15b] SCHMID B.V., BÜNTGEN U., EASTERDAY W.R. *et al.*, "The pulse of Asia: the Black Death and successive reintroductions of plague into Europe", *Proceedings National Academy of Sciences USA*, no. 112, pp. 3020–3025, 2015.

[SCH 15c] SCHRAMSKI J.R., GATTIE D.K., BROWN J.H., "Human domination of the biosphere: rapid discharge of the earth-space battery foretells the future of humankind", *Proceedings National Academy of Sciences USA*, no. 112, pp. 9511–9517, 2015.

[SCO 00] SCOW M.K., FOGG G.E., HINTON D.E., JOHNSON M.L., *Integrated Assessment of Ecosystem Health*, Lewis Publishers, Boca Raton, 2000.

[SCO 11] SCOTT J.M., *An Introduction to Evolutionary Ethics*, Wiley-Blackwell, Oxford, 2011.

[SEC 10] SECRETARIAT DE LA CONVENTION DE RAMSAR, "Lois et institutions: Compétences participatives. Mise en œuvre et renforcement de la participation des communautés locales et des populations autochtones à la gestion des zones humides", Manuels Ramsar pour l'utilisation rationnelle des zones humides, vol. 7, 4th ed., Gland, 2010.

[SEC 14] SECADES C., O'CONNOR B., BROWN C. *et al.*, Earth Observation for Biodiversity Monitoring: A Review of Current Approaches and Future Opportunities for Tracking Progress towards the Aichi Biodiversity Targets, Secretariat of the Convention on Biological Diversity, Technical Series No. 72, 183 pages, Canada, Montreal, 2014.

[SEC 16] SECONDE ASSEMBLEE DES NATIONS UNIES POUR L'ENVIRONNEMENT, Programme Des Nations Unies Pour L'environnement, Résolution 2/5. Mise en œuvre du Programme de développement durable à l'Horizon 2030, 5pp. Nairobi, 23–27 May 2016.

[SEI 12] SEILER C., BERENDONK, T.U., "Heavy metal driven co-selection of antibiotic resistance in soil and water bodies impacted by agriculture and aquaculture", *Frontiers Microbiology*, vol. 3, Article 399, p. 7, 2012.

[SEN 85] SEN A., *Commodities and Capabilities*, Elsevier, Amsterdam, 1985.

[SEN 10] SEN A., *The Idea of Justice*, Penguin, London, 2010.

[SEN 16] SENZAKI M., YAMAURA Y., FRANCIS C.D. *et al.*, Traffic noise reduces foraging efficiency in wild owls, Scientific Report, 2016;6:30602. doi:10.1038/srep30602, pp. 1–7, 2016.

[SET 10] SETO K.C., SANCHEZ-RODRIGUEZ R., FRAGKIAS M., "The new geography of contemporary urbanization and the environment", *Annual Review Environmental Resources,* no. 35, pp. 167–194, 2010.

[SET 12a] SETO K.C., GUNERALP B., HUTYRA L.R., "Global forecasts of urban expansion to 2030 and direct impacts on biodiversity and carbon pools", *Proceedings National Academy of Sciences USA*, no. 109, pp. 16083–16088, 2012.

[SET 12b] SETO K.C., REENBERG A., BOONE C.G. *et al.*, "Urban land teleconnections and sustainability", *Proceedings of the National Academy of Sciences*, no. 109, pp. 7687–7692, 2012.

[SET 13] SETO K.C., PARNELL S., ELMQVIST T., "A global outlook on urbanization", in ELMQVIST T., FRAGKIAS M., GOODNESS J. *et al.* (eds), *Urbanization, Biodiversity and Ecosystem Services: Challenges and Opportunities A Global Assessment*, London, Springer, 2013.

[SEY 16] SEYMOUR V., "The Human–Nature relationship and its impact on Health: a critical review", *Frontiers in Public Health*, vol. 4, Article 260, 2016.

[SHA 10] SHAVER L., "The right to science and culture", *Wisconsin Law Review*, no. 1, pp. 121–184, 2010.

[SHE 11] SHENOY K., CROWLEY P.H., "Endocrine disruption of male mating signals: ecological and evolutionary implications", *Functional Ecology*, no. 25, pp. 433–448, 2011.

[SHI 09] SHIRLEY S.M., KARK S., "The role of species traits and taxonomic patterns in alien bird impacts", *Global Ecology and Biogeography*, no. 18, pp. 450–459, 2009.

[SHO 14] SHORE R.F., TAGGART M.A., SMITS J. *et al.*, "Detection and drivers of exposure and effects of pharmaceuticals in higher vertebrates", *Philosophical Transactions of the Royal Society London B*, no. 369, p. 20130570, 2014.

[SIE 11] SIEMERS B.M., SCHAUB A., "Hunting at the highway: traffic noise reduces foraging efficiency in acoustic predators", *Proceedings of the Royal Society London B*, no. 278, pp. 1646–1652, 2011.

[SIL 08] SILBERGELD E.K., GRAHAM J., PRICE L.B., "Industrial food animal production, antimicrobial resistance, and human health", *Annual Review of Public Health*, no. 29, pp. 151–169, 2008.

[SIN 75] SINGER P., *Animal Liberation*, Harper Collins, New York, 1975.

[SKE 04] SKEVINGTON S.M., LOTFY M., O'CONNELL K.A., "The World Health Organization's WHOQOL-BREF Quality of Life Assessment: psychometric properties and results of the international field trial. A report from the WHOQOL Group", *Quality of Life Research*, no. 23, pp. 299–310, 2004.

[SMI 02] SMIL V., *The Earth's Biosphere. Evolution, Dynamics, and Change*, Cambridge, Massachusetts Institute of Technology, 2002.

[SMI 07] SMITH K.F., SAX D.F., GAINES S.D. *et al.*, "Globalization of human infectious disease", *Ecology*, no. 88, pp. 1903–1910, 2007.

[SMI 08] SMIL V., *Global Catastrophes and Trends: The Next Fifty Years*, Massachusetts Institute of Technology, Cambridge, 2008.

[SMI 09] SMITH N.H., HEWINSON R.G., KREMER K. *et al.*, "Myths and misconceptions: the origin and evolution of *Mycobacterium tuberculosis*", *Nature Review Microbiology*, no. 7, pp. 537–544, 2009.

[SMI 14] SMITH K.F., GOLDBERG M., ROSENTHAL S. *et al.*, "Global rise in human infectious disease outbreaks", *Journal of The Royal Society Interface*, no. 11, p. 20140950, 2014.

[SMU 12] SMULEWICZ-ZUCKER G.R., *Strangers To Nature: Animal Lives And Human Ethics*, Lexington Books, Lanham, 2012.

[SOL 71] SOLECKI M., *Shanidar: the First Flower People*, Alfred A. Knopf, New York, 1971.

[SOR 43] SORRE M., *Les Fondements de la Géographie Humaine, vol. 1, Les Fondements Biologiques: Essai d'une Ecologie de l'Homme*, Armand Colin, Paris, 1943.

[SOR 08] SOREK R., KUNIN V., HUGENHOLTZ P., "CRISPR – a widespread system that provides acquired resistance against phages in bacteria and archaea", *Nature Review Microbiology*, no. 6, pp. 181–186, 2008.

[SOR 16] SORCI G., GUIVIER E., LIPPENS C. *et al.*, "Microbes, parasites and immune diseases", in ALVERGNE A., JENKINSON C., FAURIE C. (eds), *Evolutionary Thinking in Medicine. From Research to Policy and Practice*, Springer, pp. 211–223, 2016.

[SOU 16] SOUCHET J., AUBRET F., "Revisiting the fear of snakes in children: the role of aposematic signalling", *Scientific Report*, no. 6, article 37619, 2016.

[SPA 15] SPASH C.L., ASLAKSEN J., "Re-establishing an ecological discourse in the policy debate over how to value ecosystems and biodiversity", *Journal of Environmental Management*, no. 159, pp. 245–253, 2015.

[SPI 86] SPINAGE C.A., *The Natural History of Antelopes*, Croom Helm, London, 1986.

[SPR 16] SPRINGER Y.P., HOEKMAN D., JOHNSON P.T.J. *et al.*, "Tick-, mosquito-, and rodent-borne parasite sampling designs for the National Ecological Observatory Network", *Ecosphere*, no. 7, p. e01271, 2016.

[SPU 10] SPURGIN L.G., RICHARDSON D.S., "How pathogens drive genetic diversity: MHC, mechanisms and misunderstandings", *Proceedings Royal Society London B*, no. 277, pp. 979–988, 2010.

[STE 07] STEFFEN W., CRUTZEN P.J., MCNEILL J.R., "The Anthropocene: are humans now an overwhelming force of nature?", *Ambio*, no. 36, pp. 614–621, 2007.

[STE 08] STEARNS S.C., KOELLA J.C., *Evolution in Health and Disease*, 2nd ed., Oxford University Press, New York, 2008.

[STI 09] STIRLING A.C., SCOONES I., "From risk assessment to knowledge mapping: science, precaution, and participation in disease ecology", *Ecology and Societies*, no. 14, p. 14, 2009.

[STI 10] STIRLING A.C., "Keep it complex", *Nature*, no. 468, pp. 1029–1031, 2010.

[STO 13] STOCKER T.F., QIN D., PLATTNER H.-K. *et al.* (eds), *Climate Change 2013: The Physical Science Basis. Contribution of Working Group I to the Fifth Assessment Report of the Intergovernmental Panel on Climate Change*, IPCC, Cambridge University Press, Cambridge, available at: http://www.climatechange2013.org/report/full-report/, 2013.

[STR 89] STRACHAN D.P., "Hay fever, hygiene, and household size", *British Medical Journal*, no. 299, pp. 1259–1260, 1989.

[STR 16] STRONA G., FATTORINI S., "Are generalist parasites being lost from their hosts?", *Journal of Animal Ecology*, no. 85, pp. 621–623, 2016.

[SUM 11] SUMMERS P.D., CUNNINGTON G.M., FAHRIG L., "Are the negative effects of roads on breeding birds caused by traffic noise?", *Journal of Applied Ecology*, no. 48, pp. 1527–1534, 2011.

[SUT 06] SUTHERLAND W.J., "Predicting the ecological consequences of environmental change: a review of the methods", *Journal of Applied Ecology*, no. 43, pp. 599–616, 2006.

[SUT 12] SUTHERLAND W.J., FRECKLETON R.P., "Making predictive ecology more relevant to policy makers and practitioners", *Philosophical Transactions of the Royal Society B*, no. 367, pp. 322–330, 2012.

[SWA 08] SWADDLE J.P., CALOS S.E., "Increased avian diversity is associated with lower incidence of human West Nile infection: observation of the dilution effect", *PLoS ONE*, no. 3, p. e2488, 2008.

[SWA 15] SWADDLE J.P., FRANCIS C.D., BARBER J.R. *et al.*, "A framework to assess evolutionary responses to anthropogenic light and sound", *Trends in Ecology and Evolution*, no. 30, pp. 550–560, 2015.

[TAB 13] TABASHNIK B.E., BRÉVAULT T., CARRIÈRE Y., "Insect resistance to Bt crops: lessons from the first billion acres", *Nature Biotechnology*, no. 31, pp. 510–521, 2013.

[TAI 88] TAINTER J.A., *The Collapse of Complex Societies*, Cambridge University Press, Cambridge, 1988.

[TAK 02] TAKANO T., NAKAMURA K., WATANABE M., "Urban residential environments and senior citizens' longevity in megacity areas: the importance of walkable green spaces", *Journal of Epidemiology and Community Health*, no. 56, pp. 913–918, 2002.

[TAT 06] TATEM A.J., ROGERS D.J., "Global transport networks and infectious disease spread", *Advances in Parasitology*, no. 62, pp. 293–343, 2006.

[TAY 01] TAYLOR L.H., LATHAM S.M., WOOLHOUSE M.E.J., "Risk factors for human disease emergences", *Philosophical Transaction of the Royal Society of London B*, no. 356, pp. 983–98, 2001.

[TAY 08] TAYLOR P., "Ecological integrity and human rights", in WENSTRA L., BOSSELMANN K., WESTRA R. (eds), *Reconciling Human Existence with Ecological Integrity. Science, Ethics, Economics and Law*, Earthscan, London, 2008.

[TEE 10] TEEB, *The Economics of Ecosystems and Biodiversity Ecological and Economic Foundations*, KUMAR PUSHPAM P. (ed.), Earthscan, 2010.

[TEG 04] TEGTMEIER E.M., DUFFY M.D., "External costs of agricultural production in the United States", *International Journal of Agricultural Sustainability*, no. 2, p. 12659, 2004.

[TEL 05] TELFER S., BOWN K.J., SEKULES R. *et al.*, "Disruption of a host–parasite system following the introduction of an exotic host species", *Parasitology*, no. 130, pp. 661–668, 2005.

[THE 03] THEIS N., LERDAU M., "The evolution of function in plant secondary metabolites", *International Journal of Plant Sciences*, no. 164, pp. S93–S102, 2003.

[THO 04] THOMAS C.D., CAMERON A., GREEN R.E. *et al.*, "Extinction risk from climate change", *Nature*, no. 427, pp. 145–148, 2004.

[THO 10] THORNTON P.P.K., "Livestock production: recent trends, future prospects", *Philosophical Transaction of the Royal Society of London B*, no. 365, pp. 2853–2867, 2010.

[THO 15] THOMPSON P.B., LIST M., "Ebola needs one bioethics", *Ethics, Policy and Environment*, no. 18, pp. 96–102, 2015.

[TIL 11] TILMAN D., BALZER C., HILL J. *et al.*, "Global food demand and the sustainable intensification of agriculture", *Proceedings National Academy of Sciences USA*, no. 108, pp. 20260–20264, 2011.

[TIT 14] TITTENSOR D.P., WALPOLE M., HILL S.L.L. *et al.*, "A mid-term analysis of progress toward international biodiversity targets", *Science*, no. 346, pp. 241–244, 2014.

[TOO 90] TOOBY J., COSMIDES L., "The past explains the present: emotional adaptations and the structure of ancestral environments", *Ethology and Sociobiology*, no. 11, pp. 375–424, 1990.

[TU 11] TU Y., "The discovery of artemisinin (qinghaosu) and gifts from Chinese medicine", *Nature Medicine*, no. 17, pp. 1217–1220, 2011.

[TUA 74] TUAN Y.-F., *Topophilia: A Study of Environmental Perception, Attitudes and Values*, Columbia University Press, Columbia, 1974.

[TUC 05] TUCKER G., BUBB P., DE HEER M. *et al.*, *Guidelines for Biodiversity Assessment and Monitoring for Protected Areas*, KMTNC, Kathmandu, 2005.

[TUR 14] TURNER G., Is Global Collapse Imminent, MSSI Research Paper No. 4, Melbourne Sustainable Society Institute, The University of Melbourne, 2014.

[TUR 16] TURNHOUT E., DEWULF A., HULME M., "What does policy-relevant global environmental knowledge do? The cases of climate and biodiversity", *Current Opinion in Environmental Sustainability*, no. 18, pp. 65–72, 2016.

[UDI 14] UDIKOVIC-KOLIC N., WICHMANN F., BRODERICK N.A. *et al.*, "Bloom of resident antibiotic-resistant bacteria in soil following manure fertilization", *Proceedings National Academy of Sciences USA*, no. 111, pp. 15202–15207, 2014.

[ULA 86] ULANOWICZ R.E., *Growth and Development: Ecosystems Phenomenology*, Springer-Verlag, New York, 1986.

[ULR 77] ULRICH R.S., "Visual landscape preference: a model and application", *Man-Environment Systems*, no. 7, pp. 279–293, 1977.

[ULR 83] ULRICH R.S., "Aesthetic and effective response to natural environment", in ALTMAN I., WOLHWILL J.F. (eds), *Behavior and the Natural Environment*, Plenum Press, New York, 1983.

[ULR 84] ULRICH R.S., "View from a window may influence recovery from surgery", *Science*, no. 224, pp. 420–421, 1984.

[ULR 91] ULRICH R.S., SIMONS R.F., LOSITO E. *et al.*, "Stress recovery during exposure to natural and urban environments", *Journal of Environmental Psychology*, no. 11, pp. 201–230, 1991.

[ULR 93] ULRICH R.S., "Biophilia, biophobia and natural landscapes", in KELLERT R., WILSON E.O. (eds), *The Biophilia Hypothesis*, Island Press, Washington, DC, 1993.

[UNC 17] UNCKLESS R.L., CLARK A.G., MESSER P.W., "Evolution of resistance against CRISPR/Cas9 gene drive", *Genetics*, no. 205, pp. 827–841, 2017.

[UNE 70] UNESCO, "Utilisation et conservation de la biosphère: actes de la Conférence intergouvernementale d'experts sur les bases scientifiques de l'utilisation rationnelle et de la conservation des ressources de la biosphère Paris, 4–13 September 1968", *Recherches sur les Ressources Naturelles*, no. 10, p. 305, Paris, 1970.

[UNI 14] UNITED NATIONS, World Urbanization Prospects: The 2014 Revision Highlights, Report, Department of Economic and Social Affairs, Population Division (ST/ESA/SER.A/352), United Nations, New York, 2014.

[UNI 15a] UNITED NATIONS, World Population Prospects: The 2015 Revision, Report, Department of Economic and Social Affairs, Population Division, United Nations, New York, 2015.

[UNI 15b] UNITED NATIONS DEVELOPMENT PROGRAMME, CHINA, Issue Brief: Environmental Public Interest Litigation in China, Report, Rule of Law Roundtable, p. 2, May 2015.

[UNI 17a] UNITED NATIONS, Report of the Special Rapporteur on the right to food, Human Rights Council Thirty-fourth session, Promotion and protection of all human rights, civil, political, economic, social and cultural rights, including the right to development, A/HRC/34/48, 2017.

[UNI 17b] UNITED NATIONS ENVIRONMENT PROGRAMME, The Status of Climate Change Litigation – A Global Review, Report, UNEP, 2017.

[UNI 06] UNITED NATIONS, Report of the Meeting on Indigenous Peoples and Indicators of Well-being, E/CN.e9/2006/CRP.3, Ottawa, Canada, 22–23 March, 2006.

[VAL 10] VALE P.F., LITTLE T.J., "CRISPR-mediated phage resistance and the ghost of coevolution past", *Proceedings Royal Society London B*, vol. 277, pp. 2097–2103, 2010.

[VAL 13] VALLE D., CLARK J., "Conservation efforts may increase malaria burden in the Brazilian Amazon", *PLoS ONE*, vol. 8, p. e57519, 2013.

[VAN 98] VAN BAALEN M., "Coevolution of recovery ability and virulence", *Proceedings Royal Sociedty London B*, vol. 265, pp. 317–325, 1998.

[VAN 99] VANLEEUWEN J.A., WALTNER-TOEWS D., ABERNATHY T. *et al.*, "Evolving models of human health toward an ecosystem context", *Ecosystem Health*, vol. 5, pp. 204–219, 1999.

[VAN 06] VAN VUUREN D., SALA O., PEREIRA H.M., "The future of vascular plant diversity under four global scenarios", *Ecology and Society*, vol. 11, p. 25, 2006.

[VAN 07] VAN ERT M.N., EASTERDAY W.R., HUYNH L.Y. *et al.*, "Global genetic population structure of *Bacillus anthracis*", *PLoS ONE*, vol. 2, p. e461, 2007.

[VAN 12] VANDEBROEK I., BALICK M.J., "Globalization and loss of plant knowledge: challenging the paradigm", *PLoS ONE*, vol. 7, p. e37643, 2012.

[VAN 13] VAN DE GEVEL J., GELLAW A., FADDA C., Seeds for Needs: participatory variety selection, Biodiversity International, CGIAR, Policy Brief no 3, 2013.

[VAN 14] VAN BOECKEL T.P., GANDRA S., ASHOK A. *et al.*, "Global antibiotic consumption 2000 to 2010: an analysis of national pharmaceutical sales data", *Lancet Infectious Diseases*, vol. 14, pp. 742–750, 2014.

[VAN 16] VAN DIEPEN L.T.A., FREY S.D., LANDIS E.A. *et al.*, "Fungi exposed to chronic nitrogen enrichment are less able to decay leaf litter", *Ecology*, vol. 98, pp. 1–20, 2016.

[VEE 93] VEENHOVEN R., *Happiness in nations. subjective appreciation of life in 56 nations, 1946–1992*, Erasmus University, Rotterdam, 1993.

[VEE 04] VEENHOVEN R., "World Happiness Database", available at: www.eur.nl/fsw/research/happiness, 2004.

[VEL 13] VELASQUEZ-MANOFF M., *An Epidemic of Absence: A New Way of Understanding Allergies and Autoimmune Diseases*, Scribner, New York, 2013.

[VER 88] VERMUND S.H., MACLEOD S., "Is pinworm a vanishing infection?", *American Journal Diseases of Children*, vol. 142, pp. 566–568, 1988.

[VER 02] VERBEEK P., DE WAAL F.B.M., "The primate relationship with nature: biophilia as a general pattern", in KAHN P.H., KELLERT S.R. (eds), *Children and Nature: Psychological, Sociocultural, and Evolutionary Investigations*, Massachusetts Institute of Technology, Cambridge, 2002.

[VER 16] VERWEIJ M., BOVENKERK B., "Ethical Promises and Pitfalls of OneHealth", *Health Ethics*, vol. 9, pp. 1–4, 2016.

[VIL 06] VILLALBA J.J., PROVENZA F.D., SHAW R., "Sheep self-medicate when challenged with illness-inducing foods", *Animal Behaviour*, vol. 71, pp. 1131–1139, 2006.

[VIL 07] VILLALBA J.J., PROVENZA F.D., "Self-medication and homeostatic behaviour in herbivores: learning about the benefits of nature's pharmacy", *Animal*, vol. 1, pp. 1360–1370, 2007.

[VIL 09] VILLALBA, J.J., PROVENZA, F.D., HALL, J.O. *et al.*, "Selection of tannins by sheep in response to gastro-intestinal nematode infections", *Journal of Animal Science*, vol. 88, no. 6, pp. 2189–2198, 2009.

[VIT 86] VITOUSEK P.M., EHRLICH P.R., EHRLICH A.H. *et al.*, "Human appropriation of the products of photosynthesis", *BioScience*, vol. 36, pp. 368–373, 1986.

[VON 11] VON HERTZEN L., HANSKI I., HAAHTELA T., "Natural immunity", *EMBO reports*, vol. 12, pp. 1089–1093, 2011.

[VÖR 10] VÖRÖSMARTY C.J., MCINTYRE P.B., GESSNER M.O. *et al.*, "Global threats to human water security and river biodiversity", *Nature*, vol. 467, pp. 555–561, 2010.

[VOR 14] VOREADES N., KOZIL A., WEIR T.L., "Diet and the development of the human intestinal microbiome", *Frontiers Microbiology*, vol. 5, p. 494, 2014.

[WAL 07] WALLACE K.J., "Classification of ecosystem services: problems and solutions", *Conservation Biology*, vol. 139, pp. 235–246, 2007.

[WAL 09] WALPOLE M., ALMOND R.E., BESANÇON C. *et al.*, "Tracking progress toward the 2010 Biodiversity target and beyond", *Science*, vol. 325, pp. 1503–1504, 2009.

[WAL 17] WALTERS M., SCHOLES R.J., *The GEO Handbook on Biodiversity Observation Networks*, Springer Open, 2017.

[WAM 14] WAMMERS L.J., MPAIRWE H., ELLIOTT A.M. *et al.*, "Helminth therapy or elimination: epidemiological, immunological, and clinical considerations", *Lancet Infectious Diseases*, vol. 14, pp. 1150–1162, 2014.

[WAN 06] WANG L.F., SHI Z., ZHANG S. *et al.*, "Review of bats and SARS", *Emerging Infectious Diseases*, vol. 12, pp. 1834–1840, 2006.

[WAN 16] WANG Z., PAN Q., GENDRON P. *et al.*, "CRISPR/Cas9-derived mutations both inhibit HIV-1 replication and accelerate viral escape", *Cell Reports*, vol. 15, pp. 481–489, 2016.

[WAR 97] WARREN K.J., *Ecofeminism: Women, Culture, Nature*, Indiana University Press, Indianapolis, 1997.

[WAT 16] WATERS C.N., ZALASIEWICZ J., SUMMERHAYES C. *et al.*, "The Anthropocene is functionally and stratigraphically distinct from the Holocene", *Science*, vol. 351, pp. 137–148, 2016.

[WEA 04] WEATHERALL M., MCPHERSON K., TAYLOR W. *et al.*, "Avoiding pitfalls of correlation coefficients in the assessment of measurement instruments in rehabilitation research", *Clinical Rehabilitation*, vol. 18, pp. 186–194, 2004.

[WEB 15] WEBBER B.L., RAGHU S., EDWARDS O.R., "Is CRISPR-based gene drive a biocontrol silver bullet or global conservation threat?", *Proceedings National Academy of Sciences USA*, vol. 112, pp. 10565–10567, 2015.

[WEI 04] WEINSTOCK J.V., SUMMERS R., ELLIOTT D.E., "Helminths and harmony", *Gut*, vol. 53, pp. 7–9, 2004.

[WEI 13a] WEIGERS VITULLO M., WYNDHAM J., Defining the Right to Enjoy the Benefits of Scientific Progress, American Scientists' Perspectives, Report, American Association for the Advancement of Science, 2013.

[WEI 13b] WEILL M., "La mauvaise gestion des insecticides un danger pour la santé publique et l'environnement", in MORAND S., PIPIEN G. (eds), *Notre Santé et la Biodiversité*, Buchet-Chastel, Arles, 2013.

[WHA 08] WHALON M.E., MOTA-SANCHEZ D., HOLLINGWORTH R.M., "Analysis of global pesticide resistance in arthropods", in WHALON M.E., MOTA-SANCHEZ D., HOLLINGWORTH R.M. (eds), *Global Pesticide Resistance in Arthropods*, CAB International, Wallingford, 2008.

[WHI 09] WHITT L., *Science, Colonialism, and Indigenous Peoples. The Cultural Politics of Law and Knowledge*, Cambridge University Press, Cambridge, 2009.

[WHI 15] WHITMEE S., HAINES A., BEYRER C. *et al.*, "Safeguarding human health in the Anthropocene epoch: report of The Rockefeller Foundation–Lancet Commission on planetary health", *Lancet*, vol. 386, pp. 1973–2028, 2015.

[WHO 86] WORLD HEALTH ORGANIZATION, Charte d'Ottawa pour la promotion de la santé, World Health Organization, Geneva, 1986.

[WHO 93] WORLD HEALTH ORGANIZATION, WHO Global Strategy for Health and Environment, WHO/EHE/93.2, Report, Geneva, 1993.

[WHO 93a] WORLD HEALTH ORGANIZATION SEARO, Regional Strategic Plan for Health and Environment for the WHO South-East Asia Region, Report, World Health Organization, Geneva, 1993.

[WHO 93b] WORLD HEALTH ORGANIZATION WPRO, Regional Strategy on Health and Environment, including follow-up of the United Nations Conference on Environment and Development, Report, WPR RC44/13, World Health Organization, Geneva, 1993.

[WHO 98] WORLD HEALTH ORGANIZATION QOL (The World Health Organization Quality of Life Assessment), "The World Health Organization Quality of Life Assessment: development and general psychometric properties", *Social Sciences and Medicine*, vol. 46, pp. 1569–1585, 1998.

[WHO 05] WORLD HEALTH ORGANIZATION SEARO, Combating Emerging Infectious Diseases in the South-East Asia Region (SEA-CD-139), Report, World Health Organization, Geneva, 2005.

[WHO 06] WORLD HEALTH ORGANIZATION, *WHO International Health Regulations*, 2nd ed., WHO, Geneva, 2006.

[WHO 07] WORLD HEALTH ORGANIZATION, A Safer Future: Global Public Health Security in the 21st Century, Report, 2007.

[WHO 09] WORLD HEALTH ORGANIZATION, International Health Regulations, 2005. Brief introduction to application in the context of national legislation, WHO/HSE/IHR/2009, Report, Geneva, 2009.

[WHO 11] WORLD HEALTH ORGANIZATION, The World Traditional Medicines Situation, in Traditional medicines: Global Situation, Issues and Challenges, no. 3, pp. 1–14, World Health Organization, Report, Geneva, 2011.

[WHO 14a] WORLD HEALTH ORGANIZATION, "Antimalarial Drug Resistance", World Health Organization, Geneva, available at: http://www.who.int/malaria/areas/drug_resistance/overview/en/, 2014.

[WHO 14b] WORLD HEALTH ORGANIZATION, Antimicrobial Resistance: Global Report on Surveillance, Report, World Health Organization, Geneva, Switzerland, 2014.

[WHO 15a] WORLD HEALTH ORGANIZATION, World Malaria Report 2015, Report, World Health Organization, Geneva, 2015.

[WHO 15b] WORLD HEALTH ORGANIZATION – SCBD, Connecting Global Priorities: Biodiversity and Human Health: A State Of Knowledge Review, Report, World Health Organization and Secretariat of the Convention on Biological Diversity, 2015.

[WIL 75] WILSON E.O., *Sociobiology: The New Synthesis*, Harvard University Press, Cambridge, 1975.

[WIL 84] WILSON E.O., *Biophilia: The Human Bond with Other Species*, Harvard University Press, Cambridge, 1984.

[WIL 99] WILLIAMS A., *Therapeutic Landscapes: The Dynamics between Place and Wellness*, University Press of America, Lanham, 1999.

[WIL 04] WILLCOX M., BODEKER G., RASOANAIVO P., *Traditional Medicinal Plants and Malaria*, CRC Press, Florida, 2004.

[WIL 05a] WILCOX B.A., COLWELL R.R., "Emerging and reemerging infectious diseases: biocomplexity as an interdisciplinary paradigm", *EcoHealth*, vol. 2, pp. 244–257, 2005.

[WIL 05b] WILCOX B.A., GUBLER D.J., "Disease ecology and the global emergence of zoonotic pathogens", *Environmental Health Preventive Medicine*, vol. 10, pp. 263–272, 2005.

[WIL 10] WILLIAMS A., "Therapeutic Landscapes as Health Promoting Places", in BROWN T., MCLAFFERTY S., MOON G. (eds), *A Companion to Health and Medical Geography*, Blackwell, Oxford, 2010.

[WIN 11] WINDHAGER S., ATZWANGER K., BOOKSTEIN F.L. *et al.*, "Fish in a mall aquarium–An ethological investigation of biophilia", *Landscape and Urban Planning*, vol. 99, pp. 23–30, 2011.

[WIR 05] WIRTH T., MEYER A., ACHTMAN M., "Deciphering host migrations and origins by means of their microbes", *Molecular Ecology*, vol. 14, pp. 3289–3306, 2005.

[WOL 07] WOLFE N. D., PANOSIAN DUNAVAN C., DIAMOND J., "Origins of major human infectious diseases", *Nature*, no. 447, pp. 279–283, 2007.

[WOL 14] WOLF M., "Is there really such a thing as 'One Health'? Thinking about a more than human world from the perspective of cultural anthropology", *Social Science and Medicine*, vol. 129, pp. 5–11, 2014.

[WOO 05] WOOLHOUSE M.E.J., GOWTAGE-SEQUERIA S., "Host range and emerging and reemerging pathogens", *Emerging Infectious Diseases*, vol. 11, pp. 1842–1847, 2005.

[WOO 07] WOOLHOUSE M., GAUNT E., "Ecological origins of novel human pathogens", *Critical Reviews in Microbiology*, vol. 33, pp. 231–242, 2007.

[WRA 95] WRANGHAM R.W., "Relationship of chimpanzee leaf-swallowing to a tapeworm infection", *American Journal of Primatology*, vol. 37, pp. 297–303, 1995.

[WRI 83] WRIGHT D.H., "Species-energy theory: an extension of species-area theory", *Oikos*, vol. 41, pp. 496–506, 1983.

[WRI 90] WRIGHT D.H., "Human impacts on energy flow through natural ecosystems, and implications for species endangerment", *Ambio*, vol. 19, pp. 189–194, 1990.

[WRI 00] WRIGLEY S.K., HAYES M.A., THOMAS R. *et al.*, *Biodiversity: A Source of New Leads for the Pharmaceutical and Agochemical Industries*, The Royal Society of Chemistry, Cambridge, 2000.

[WWF 10] WWF (WORLD WILD LIFE FUND), Vital sites: The contribution of protected areas to human health, Report, 2010.

[YAN 10] YANG A.C., HUANG N.E., PENG C.K. *et al.*, "Do seasons have an influence on the incidence of depression? The use of an internet search engine query data as a proxy of human affect", *PLoS ONE*, vol. 5, pp. 1–7, 2010.

[ZEL 14] ZELENSKI J.M., NISBET E.K., "Happiness and feeling connected: the distinct role of nature relatedness", *Environment and Behavior*, vol. 46, pp. 3–23, 2014.

[ZHA 05] ZHAO J.-Z., CAO C., COLLINS H.L. *et al.*, "Concurrent use of transgenic plants expressing a single and two Bacillus thuringiensis genes speeds insect adaptation to pyramided plants", *Proceedings National Academy of Science USA*, vol. 102, pp. 8426–8430, 2005.

[ZHA 07] ZHANG W., RICKETTS T.H., KREMEN C. *et al.*, "Ecosystem services and dis-services to agriculture", *Ecological Economics*, vol. 64, pp. 253–260, 2007.

[ZIE 05] ZIERSCH A.M., BAUM F.E., MACDOUGALL C. *et al.*, "Neighbourhood life and social capital: the implications for health", *Social Sciences and Medicine*, vol. 60, pp. 71–86, 2005.

[ZIN 12] ZINSSTAG J., MACKENZIE J.S., JEGGO M. *et al.,* "Mainstreaming one health", *Ecohealth*, vol. 9, pp. 107–110, 2012.

[ZUR 14] ZUREK L., GHOSH A., "Insects represent a link between food animal farms and the urban environment for antibiotic resistance traits", *Applied Environmental Microbiology*, vol. 80, pp. 3562–3567, 2014.

[ZYL 13] ZYLBERMAN P., *Tempêtes Microbiennes. Essai sur la Politique de Sécurité Sanitaire dans le Monde Transatlantique*, Gallimard, Paris, 2013.

[ZYL 16] ZYLBERMAN P., "L'avenir, 'cible mouvante'. Les États-Unis, le risque NRBC et la méthode des scénarios (1995–2008)", in MORAND M., FIGUIE M. (eds), *Emergence de Maladies Infectieuses Émergentes. Risques et Enjeux de Société*, Quae, Versailles, 2016.

Index

A, B, C

D, E, F

G, H, I

L, M, N

O, P, R

S, T, U

V, W, Z

FSC
www.fsc.org
MIX
Papier | Fördert gute Waldnutzung
FSC® C083411

Kolibri 360 GmbH
Ferdinand-Jühlke-Straße 7
99095 Erfurt, Deutschland
produktsicherheit@kolibri360.de